B.L. SAUNDERS

SEVEN HALF-MILES from HOME

SEVEN HALF-MILES from HOME

Notes of a Wind River Naturalist

Mary Back

Johnson Books: Boulder

Johnson Books: Boulder

©Copyright 1985 by Mary Back

Cover design by Joyce Rossi
Cover art: *Snowy Egret* by Mary Back
Illustrations by Mary Back
Maps by Michael McNierney

ISBN 0-933472-90-0

LCCCN 85-081264

Printed in the United States of America by
Johnson Publishing Company
1880 South 57th Court
Boulder, Colorado 80301

Contents

Introduction 1
As the Land Lies 3
Seven Walks 7
The Communities 13
Wind River 15
Sloughs and Springs 43
Fences 57
Thickets 71
The Woods 111
Houses, Yards, and Gardens 127
The Desert 151
Bibliography 185

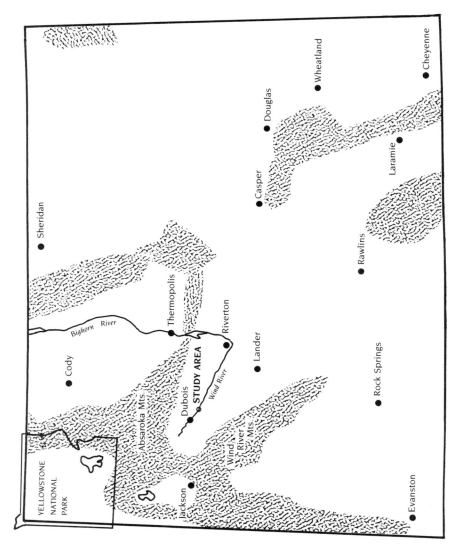

Introduction

SEVEN HALF-MILES FROM HOME is a revelation of the contrast between the casual eye and the eye of an artist. Mary Back is an artist; she sees her world with an artist's eye. Because her doctor told her she should walk a mile every morning she naturally made an artist's design from his prescription and proceeded to walk the seven pieces of that design with eyes and ears and tender heart open, most of the time with a dog companion who she feels helped her see more animal life and frightened none.

The result of this scheme is a treasury of information on one "round" mile of desert, woodland, badlands, river, and fields—information usable on countless other miles in the Rocky Mountain West and here served up to us with such humor, perception, and whimsicality that the reading is a delight.

In order to give us the full story of her walks, Mary has outlined the different kinds of habitat; each has its story but they all overlap. She says: "In this study, the constant complication is that each of the creatures (plant and animal) that I encounter is a free agent, an individual. He hasn't read the nature books to find out where he is supposed to be and who he is supposed to be and how he is supposed to act. If he had, or if he were an inert substance instead of a living child of God, I'd have nothing to write about. It would have been said already."

But it has not all been said already, and when the reading is done, here is a book to reread for pleasure in the sparkling little anecdotes and to refer to often for information on every aspect of that small but amazingly large world which is Mary's home.

Joe and Mary Back are dear friends of mine. When they lived on the highway my husband and I used to stop whenever we were on the road somewhere east of Dubois "just to say hello for a minute." Two hours later we would be strong-mindedly making ourselves climb into the car again. So I hope we may someday have the story of Mary and Joe. But here in the meantime you will come to know Mary, and she will show you how rich life can be on one "round" mile of Wyoming, which to the casual eye looks rugged, slightly interesting, very dry, and almost devoid of active life. Day by day and hour by hour Mary found a rich and varied network of life forms acting and interacting in harmonious complexity, and from her quiet observation she gives us of herself and her philosophy, and we are enriched.

Margaret E. Murie
Moose, Wyoming

As the Land Lies

~~~~~~~~~~~~~

THE SNOW WAS DARK BLUE, embroidered with black sagebrush. Above it the sky was pink and gold. The border collie, Buttons, nosed my hand gently.

Against the sky rose a pair of ears, triangular upstanding ears, like a husky's but bigger. Under them emerged the body of a coyote, like a magical emanation from the sagebrush and rocks. Silently, another appeared beside it. They were coming toward us in the growing dawnlight, graceful as moonbeams. Buttons exploded like a released spring and shot toward them, tearing the air with yelping. They retreated, light as down, easily distancing the pup and looking back over their shoulders. In no time at all the three were out of sight, the yelping growing faint. Terror gripped me. I'd heard many stories of the way coyotes will lure dogs away, then ambush them and cut their throats in the spirit of good clean fun. And these two looked so superbly sure of themselves.

The yelping stopped. I called frantically as I puffed and panted along the fresh tracks. "Buttons! Come! Here, pup! Buttons!" When she came, full on the run, I was faint with relief. Arms around each other, we sat down, gasping, on a flat snowy rock. Our hearts slowed down as we watched our valley brighten into a November day.

South, over my left shoulder, I know is the Wind River Range. I know it rears up more than 13,000 feet into the Wyoming sky. But I can't see it; from here it is blocked from view by Whiskey Mountain, 11,000 feet high but just one of the foothills.

To the north rise the gray cliffs of the Absarokas, twenty miles away and 12,000 feet tall, with their skirts ravelling out towards us into the ragged red ruffles of the Wind River Badlands.

Buttons and I are up on a glacial ridge at the edge of an old, dry lake bed, looking west from tumbled boulders of the moraine that once dammed the valley and made the lake. Jakey's Fork Glacier came northward behind us long ago in a buckled flat ribbon of ice, pulling and churning masses of dirt and rocks along its edges to make these hills. Once it was fifteen miles long, half a mile wide, and a quarter mile thick. Once it pushed so hard against the Badlands that it dug out a big scallop with a vertical red wall 500 feet high. Wind River flows against it and has cut its way through the thick moraine dam. Now, Jakey's Fork is a bright little mountain stream about two jumps wide, rippling its way over the boulders dropped from the glacier's belly. Now, where blue ice used to be, are clear pools where rainbow trout and brookies live.

How long ago was this. I wonder. I have read that the last advance of the Pleistocene ice in the Middle West was around 10,000 years ago. But here? In space we are only twenty miles removed from the Ice Age right now. That distance to the south

AS THE LAND LIES

are Gannett Glacier and the big Dinwoody Ice Field. Jakey's Fork Glacier is gone, but only just. Where it headed, above Simpson Lake and only a dozen miles away, there are still permanent snowfields. Five hundred years ago? A thousand years? Two thousand? I wonder.

A chipmunk chatters. Buttons takes off in a scramble of black fur and digs frantically—white snow and white sand under white paws—beneath a quartzite boulder. Somewhere underground the chipmunk still taunts. Buttons has quite forgotten the coyotes. I turn my back on her and look west up Wind River Valley.

The glacial lake bed is a couple miles long and about half a mile wide. It slopes gently from the gravelly south bank and the sculptured red clay north bank, down to the shallow stony channel of meandering Wind River, about seventy-five feet wide. In this lake bed plain, faced with clay and sand over older gravels, the channel has swung its course to right and left—here depositing black silt to enrich hay land, there cutting a vertical wall in red clay to delight kingfishers and bank swallows, and in two places cutting across oxbows to leave swampy crescents of sloughs.

Farther west, blue with the distance, I can see the ragged Continental Divide, thirty miles away, where the river begins at 9500 feet above sea level. Wind River Lake shines up there, I know, under the tall gray cliffs of Brooks Mountain. I can see, close to Brooks, the flattened cone of Lava Mountain, an old volcano on the Divide. Out of that reddish cone have burst the deeply pocked black rocks I can find in the bed of the river and on the nearby benches. Connecting these two peaks with the highlands north and south of me are two sweeping arms: the south arm gently undulating and heavily clothed with evergreens, the north arm rugged, ragged, naked, with tremendous crags and violently jagged passes.

This is the lay of my land. In its middle are a few acres mine and Joe's by deed and record. The rest are mine to explore, study, and love, by the charity of my neighbors and the orders of my doctor.

Bless the doctor! "It would help your circulation," he said in March of 1963, "if you would walk a mile every day before breakfast".

"Oh, but I can't," cries my conscience, "I haven't *time*." "Oh, but you must!" answers my other self sternly. "Listen to what the doctor says. Don't let those varicose veins clog up." "Wel-l-l," murmurs my conscience, weakening, "that's just what I always wanted to do anyway." "That's a girl!" approves my other self, following up her advantage. "Remember your battles with cabin fever? What's keeping you housebound is that impossible schedule of housework, teaching art, working on your own art jobs, visiting with the dudes in the studio, helping Joe with his sculpture reproductions, doing the bookkeeping and correspondence and advertising, and trying to do some church work too. Remember how you looked out the windows, then turned your eyes away, till you couldn't stand it any longer? Remember how you rushed outdoors and just screamed awhile into the wind? And then took a whole day off and tried to climb all the mountains at once, and got so tired you couldn't work for another several days? Go ahead, try this mile-before-breakfast. Maybe you won't have to give in to a wear-yourself-out-every-once-in-a-while."

Friends ask me, puzzled, "Where do you walk to? Just up and down the highway?" Astonished and just as puzzled, I answer, "Why, every direction except up the highway and down the highway."

For some time the walks were exploratory. I took the precaution of asking permission of my neighbors to cross their lands. No trouble. They were all amused by that particular medical prescription. Gradually a pattern evolved. Considering home to be the center of a circle a mile across, there came to be seven different walks, one for each day of the week, like seven different crooked clock hands aimed at seven different hours. (See the map) After a while each walk developed a strong personality and program of its own. I became an addict. I am worse than an addicted televiewer. I dare not miss a program. If I do, something will have happened and I won't know it. The circulation got better, too.

Seven Walks

MOST OF THE YEAR the walks are indeed before breakfast, just as the doctor ordered. For a few weeks around the winter solstice—from say mid-November into February—it's too dark before breakfast to see anything; those weeks I eat first and take off at dawn. As soon as dawn comes by 6:30, I start off then and eat when I get back.

Weather doesn't really matter. It can hardly ever be severe enough to hurt me in a mere mile stretch. Raincoat, plastic head scarf, and rubber boots take care of wet days; heavy wool pants, two pairs of wool socks, overshoes, padded jacket with hood take care of cold ones; an added scarf across my face takes care of high wind with blinding snow, and I walk within sight of a fence. My greatest safeguard is the blessed knowledge that if I'm more than fifteen minutes late, Joe will come looking for me. So what can really go wrong?

The Monday walk is northeast to east, or about two to three on a clock diagram. I go down Wind River on its south bank to the Motel Slough, around the Slough Woods, across the Bog Knots, behind the Red Rock Motel, and home by the road past the bridge. This takes me through several life zones or ecological communities: the live stream, with its bars, beaches, and islands; the slough; two springs, and a cattail swamp; bottomland cottonwood forest; heavy streamside brush; the log bridge; a brush-grown dike or levee; a single lane dirt road; and some houses.

Tuesday is the southeast day, or four to five on the clock diagram. I cross the Red Rock hayfield, cross the Red Rock pasture south of the Motel Slouth, go on east to the New Slough, across the gravel flats to the New Channel, then south across the highway into the Jakey's Fork moraine, then northwestward home across

SEVEN HALF-MILES FROM HOME

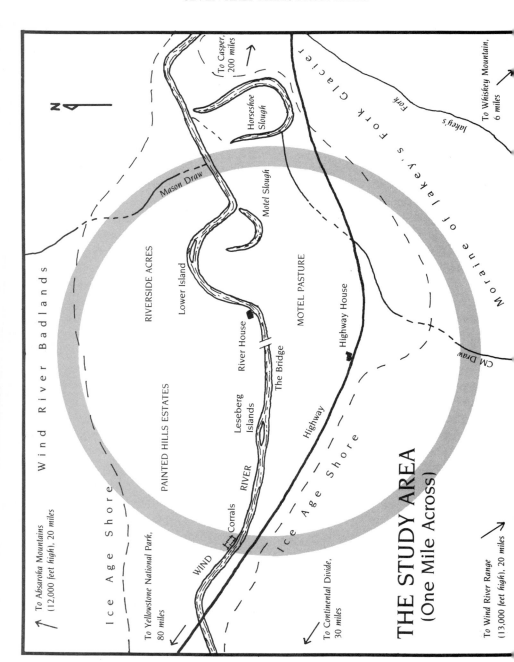

CM draw. The ecological communities are meadowland; fence rows; willow thickets; grazed pasture; gravel and mud flats; a new island covered with Cottonwoods; the highway and its edges; moraine hills spotted with sagebrush; sage flats; and the lawn, flower borders, brushy spots, and gravel parking space of home.

Wednesdays I go straight south (six o'clock on the clock diagram). I follow the section-line fence up CM Draw, climb steeply out of the draw up a moraine hill to the section corner (a pile of stones high on the east bank of the draw) back north down CM Ravine and through the Pole Gate, up the west bank of CM Draw, across the flat bench, through the gravel pit, down Gravel Pit Draw, and across the highway to home. This is the dry walk. Except close to our river-house, it's through only the sagebrush community, the walk with the widest views and the fewest birds. Just the same, I'm likely to be breathless with expectation, because here may be the most dramatic confrontations: like the two coyotes; like a rock wren on a boulder against the sky, tilting his head back and singing; like ten mule deer running among cabin-sized granite rocks; like a pair of golden eagles sitting shoulder to shoulder on a block of granite; a raven circling low over my head and exchanging croaking remarks with me; or a badger digging himself in under a sagebrush, facing an excited Buttons, answering her wild yelps with snarls of hair raising ferocity.

Thursday is southwest, or seven to nine o'clock on the diagram. I cross the highway, climb to the Low Bench, wander westward as far as I like (maybe to the steep canyon opposite Leseberg's bridge), down the canyon and northeast to the cattle pass under the highway to the water-gap at the river, eastish along the river bank to the bridge, then home. This is a wildly varied walk, with the lives of different ecological communities absurdly mixed. I can walk along the desert edge of the bench and look down on great blue herons fishing in the river and on California gulls beating up and downstream above it. I can study a water ouzel at close range as he walks about under water eating snails, while a raven sails over from the dry bench, lights in a cottonwood tree above me, and makes guttural comments. In the course of this walk I study the ecology of the sagebrush flat, of piles of boulders, of canyons just beginning, of rancher's corrals, of the live stream

with its brushy banks, of the bridge, the fence-row, and finally the cottonwood forest of home.

Friday is west and northwest, nine to eleven o'clock on the clock diagram. The ecological communities are the live stream and its brushy banks, two cottonwood copses, two islands with narrow shoreside water channels, the wide weed-patch spotted with barren alkali disks, the dirt road and fence row, and the grass and rabbitbrush of our open space.

Saturday, eleven to three o'clock on the diagram, is straight north to the Badlands, along and near the section-line fence. On this trip I look for horned larks and meadowlarks and a variety of sparrows over the flats; a colony of prairie dogs; hawks, ravens, and swallows in the cliffs; chickadees, nuthatches, vireos, and warblers in the cottonwood bottom; and ducks, kingfishers, sandpipers, and dippers; by and in the stream.

Sunday is the day for concentrating on our own acreage: the dirt road and the bridge; the open land, with turf, rye grass, and rabbitbrush; the buildings, with jutting logs, vines, fences, and flower beds; the cottonwood forest, with brush piles and undergrowth; and the river and its banks.

This is an embarrassment of riches. I can never get around to seeing it all.

Each walk is further complicated by the weather and by the changing seasons. Even if the route were just the same, which it isn't, the change in weather and seasons makes each walk a brand new thing. It's new, yet companionably old, as day by day and week by week I greet old friends and they greet me.

Most of the old friends, and the new ones, too, are remarkably closely tied to one kind of environment. After a while I began learning where and how to look for each one. A few species may turn up anywhere, and quite a few are at home in more than one life-zone. And even the most stay-at-home individual may find himself in a most strange location. I should be astonished to find a water ouzel on a Badlands cliff, or a horned toad on a gravel bar in the river, or a great blue heron in CM Ravine, or fifteen rosy finches on my windowsill. But things like this do happen. They add color and flavor and texture to the mystery of life.

To bring this embarrassment of riches into order without diminishing the joy of it is the purpose of this study. (What is it then, Mary Back, that you hope to accomplish? Why, Conscience, I hope to get numbers of people like me into a state of frantic curiosity about the walkable country around their own homes. I hope to get them as excited as I am by the discovery that all life is one thing, and that each of us is part of it. We are part of all life, and all life is part of us. We have a right to share the excitement of a birch tree when the new sap comes running up its stems, and a right to feel flattered when the bald eagle hovers low above us, looking us over with fierce yellow eyes, considering us as one equal to another.)

The key to this unity in diversity of life is the arrangement in ecological communities. So that will be my pattern. I'll study the communities in turn, first giving a summary of them all, from the river outward and upward. Then I'll have a chapter for each one in turn: the lay of the land, its geology, the plant life, the animal life, unusual experiences, and the mysteries that capture the imagination.

The Communities

I'LL ADMIT RIGHT NOW that I am using the phrase "life zone" with more freedom than scientific precision. The term is supposed to refer to a whole climatic region. In this sense, my study area is entirely in the Upper Sonoran Zone, whose most prominent feature is sagebrush desert. Four other climatic life zones are in plain sight on the mountains to north, south, and west: the Transition (lower fringe of timber), Canadian (heavy forests), Hudsonian (upper fringes of timber), and the Arctic (above timberline).

The little "life zones" I encounter on my walks are something else. The only other term I know for them is "ecological communities," a limited environment that attracts certain forms of life, so that when you find some of the factors, you usually find the rest of them—a cozy group of living beings, generally unlike except that they like to live together, recognizing each other as companionable neighbors.

Human communities are different structures. On the plus side they reflect the adaptiveness of the human animal, in that his communities may be superimposed on any climatic life zone, or on any small ecological community. On the minus side they reflect the arrogance of the human soul, in that humans generally consider themselves as creatures apart from the rest of life, masters and not neighbors; or if they can't be masters, then enemies.

Many humans dislike the position of masters and enemies, feel it's a false one, wish to break through into the friendlier relation of neighbors. Whatever their faith in immortal souls or resurrected bodies, they know that at death their bodies go back into the same reservoir of life from which they came and that after what corresponds to a chrysalis period, they come forth again.

The great joy of a study like this is conscious immersion in the body of life; swimming in it, on an equal basis with the other forms of life who are all part of it, too; learning that in life as well as death we humans are all part of the same body as all the rest of living things. Within the body of life in my mile-across area, I find these seven communities, or life zones:
- (1) the living stream, with its beaches and gravel bars
- (2) swampy places—two sloughs, with ponds, bog knots, cattail areas, and wet meadows
- (3) fence rows
- (4) thickets
- (5) forests
- (6) yards and gardens
- (7) the desert, including the foothills of Sheep Ridge and the Wind River badlands.

Then there are other life zones that have to do with this study but are too far off for my daily walks: they are the ones of the high mountains. Starting from the top, here is a list of them:
- (1) "bare rocks," cliffs, boulders (they're never really bare)
- (2) snowfields and glaciers
- (3) timberline forests of fir and stunted white-barked pine
- (4) birches, alders, and willows
- (5) fast mountain streams and waterfalls
- (6) mountain lakes
- (7) aspens
- (8) lodgepole forests
- (9) mountain meadows or "parks"
- (10) Douglas fir slopes
- (11) spruce jungles
- (12) lower timberline fringes of limber pine and juniper.

I can see these far off from the flats of my old lake bed. Every now and then—to picnic or backpack, to get firewood, to show the country to a visitor, to go hunting (for deer, elk, or rocks), to paint, or just to go—I get physically into them. Here are elk, moose, bear, and deer; "camp robbers" and "pinon squawkers;" golden-mantled ground squirrels; hoary marmots; conies and rosy finches. Sometimes some of these make expeditions into "my" country, and I shall enjoy telling about them.

Wind River

MY MILE-CIRCLE IS COMPLETELY THE CHILD of the river. In this area there is not, on the surface, one bit of "country rock." Every bit has been transported to its present spot by Wind River and its tributaries.

Not far off there is solid rock. Just out of my walking reach, a mile west, the creamy-tan Nugget Sandstone of Cretaceous age rises up in a long southwest slant from under the gravels of my study area and the red clays of the Wind River Badlands. The river has carved this into cliffs on its south side behind Louisiana-Pacific's mill. On its north side is some of the soft greenish sandstone and limestone they call the "Upper Sundance," lying just under the Badlands.

Those Wind River Badlands are the nearest thing to "country rock" in my study circle. They are red clays, partly hardened red siltstones, crumbly tan conglomerates, and whitish limy marls, standing in ragged cliffs walling in the old lake bed on its north edge. They are Eocene in age, several millions of years older than the Ice Age moraine and the clays and gravels of the lake bed, but younger than the "Upper Sundance" underneath them. They were put here by the ancestral Wind River and are made up of debris from the ancestral Wind River Mountains.

Younger than the Badlands but older than the river and the lake gravels are the volcanic layers of the Absarokas, Oligocene in age. They are not now in my study area; they make the great grey-violet mountain walls a few miles north. But once, and during the long lifetime of this same Wind River, they filled the whole valley, thousands of feet above where our heads are now. The river has cut them all away. It has cut steeply downward, dissecting away the lavas, ashes, and volcanic breccias of the Absarokas. It is still working on this project, though out of my area of study; but remains of the dissection are to be found in the gravel bars of my study-mile. Rock hunters pick up agatized wood here, evidence of the forests overwhelmed by volcanic deposits, and evidence of Wind River's transporting power.

For reasons I don't know, and at intervals I've never heard explained, the whole land pulsates, like deep breaths drawn over hundreds of years. During periods of uplift the rejuvenated river cuts fast. During sometimes long periods of subsidence the river spreads lazily out into lakes and sloughs, into which are gradually accumulated gravels from the yearly floods from the higher lands. Sometimes these deposits are hundreds of feet thick, ready to be cut down later into benches, draws, and canyons.

Part of Wind River's present project is the sculpturing of the Badlands. Its tributaries, partnered with storms and wind, have cut the soft striped rocks into intricate forms and patterns.

So the earth of my study-mile is made up of the whittlings of the river, carved by the river before, during, and after the Ice Age, brought down from the mountains by floods, dropped helter-skelter, then further mixed and ground up by storm after storm for thousands of years.

Within the last few hundreds of thousands of years the river has been cutting down, not building up, so the land must be in a rising time. It has cut through enormous thicknesses of gravel it had built up during ages before that. The record of this cutting is a series of terraces, each level showing where the river paused awhile, till a new uplift of the land set it to work again. I think there are six of these terraces around, both outside and inside my study-circle. The highest is the top of Table Mountain, 1300 feet above the present river, six miles northeast of our house and way up above the Badlands. I think it is the same level as Sheep Ridge, the high flat above Jakey's Fork Canyon south of us, where the TV relay tower is and where the mountain sheep stay in winter. If I stand on the roof-steep slope above the Pole Gate where CM Ravine opens out into CM Draw, I can see the next four lower terraces: (1) one long continuous skyline to the south, marking the edge of the "high bench" (I can't see from here, but I know that on south from that edge it's a long gentle rise for three miles to the TV tower); (2) and (3) then all the draws below the "high bench" show two humps in their divides, humps that from this point of view line up beautifully in elevation, indicating two levels later than the "high bench"; below me stretches (4) the "low bench", a quarter of a mile wide, reaching from the mouths of the draws that drain from the "high bench" to the very steep slope (a loose masonry wall of great boulders) going down to the highway and the river bottom. The river bottom itself is two levels, (5) the flats that represent the old lake bed, now turned into residential developments, and (6) the inner channel of the river, cut since lake bed days. This channel is the present theater of action of the force that is our river.

The river enters my mile-circle in the middle of its west side, where the Leseberg corrals stand on the north bank and the cattle pass comes under the highway on the south. It flows east and a little south, gently rippling past two gnarled old prostrate cedars and some large cottonwoods on the north bank and a thicket of birch, willow and silverberry on the south, to Leseberg's Islands about half-way to the bridge. There it divides into three parts. The south stream carries the most water and goes fastest, talking with a loud voice as it burbles past big boulders. The middle stream is quiet, curving northeast, then southeast, to surround an acre-sized island. It has reaches of pools and shallows and is walled with willows that arch over its water. Sandpipers nest here, mallards rest from the current, even the hardy dipper sometimes likes to be quiet here. The third channel, the north one, is temporary, carrying flood water sometimes during June. It cuts a big bite northward into Painted Hills flat. During its short period of life a piece of river rushes headlong and dirty down its grassy channel, dropping sand, mud, and little sticks among the willow shoots, taking on more sand, mud, and sticks where it is cutting into the field, dropping its load again into the quiet middle stream. The two are building a bar across the mouth of the middle stream.

Upstream from the island is a gathering place for mallards. There are shallows, with many rounded cobbles sticking out of the water, and long wavering strings of green algae trailing between them. There must be considerable mallard food, for they love to tip up here.

At the foot of the island the rapids invite the mergansers to play. They enjoy coasting fast between the rocks, flying back to the top, and repeating. The pool below the rapids is a good place for fishing. One morning I watched three red-breasted mergansers fighting there, flying at each other with spray flying. One threw in the air a fish about six inches long, then caught it and swallowed it head first. That settled it—the fight was over.

The log bridge is the mid-point of my study area. It is the link that allows a dirt road to lead from the highway to the gravel pits up Mason Draw to the north. Much of the concrete built into the town of Dubois came from these gravel pits, via heavy trucks across this bridge. The gravel traffic is mostly choked off now by relocated roads that make the hauling harder. But the bridge is now the connection between the highway and fifteen homes on the north side of the river, and ours is one of them. It is supported in the middle by a log crib filled with tons of flat hunks of red rock. The dippers may sleep here in the winter—I often see them flying out from under the bridge floor at dawn. In summer cliff swallows and robins nest under this floor and against the crib. This is a fine vantage point. Upstream and downstream, straight down and up above, and both banks, all bring their rewards. Often the biggest share of my daily wildlife notes are made on the bridge.

Below the bridge the river continues its steady rippling for a hundred yards or so eastward, then sends a push of fast water against log cribs on the south bank and rebounds to make a swing to the northeast for a quarter of a mile. It flows between a cottonwood forest on the north bank (where our river-house is) and the Red Rock meadow on the south. Just past the meadow is the Red Rock Ranch Motel, then cottonwood forest on both sides. Right where there's woodland on both sides, its channel is broken by the Motel Islands (one to several low gravel bars that disappear at flood times and on reappearance have changed their shape and position).

Beyond them, beside Bob Tripp's house and yard, is a wide pool, much enjoyed by mallards and Barrows goldeneyes. It is at the head of Lower Island where the river is right now making its most rapid changes. Lower Island was begun only about fifteen years ago, when the June flood left near the middle of the river a derelict cottonwood trunk, its swirl of tangled roots upstream. Year by year the snag caught mud and gravel. Willows and grass began to grow. The island became solid land, higher than all but the highest floods.

Where the fast water cuts deepest into the sandy bank, it is thrown back at right angles, making the biggest direction change in my area of study, from northeast to southeast. There is something dramatic about this change that is associated in my mind with unusual wildlife activity: an osprey plunging for fish; a great blue heron waiting in the shadow; a whole family of red-necked grebes. I saw these last in early October, before sunrise, with ghosts of light mist over the water. Between scarves of mist, a big-headed adult and four look-alikes not quite as large slid in a solemn, nodding procession down the north channel away from the island and on downstream. I looked hard for field marks, for I knew these must be rare birds here. They were close to me, and right at hand in the glasses: almost as large as Barrow's goldeneyes, big heads and slender necks, the leader with a white crescent under her jaw and up behind her ear. As they came into the main stream, they did the trick that proved them grebes—all five sank down in the water till their backs were out of sight, and there was just a procession of heads and necks.

The river goes on, southeast, to where the Motel Slough sends in a small stream from the south. It makes a wide curve here, northeast again for a way, cutting vertical red banks on the north, leaving low gravel flats on the south, another hundred yards or so, past the mouth of Mason Draw, a channel from the Badlands that carries water only during and just after a big storm. Finally, here it turns east in a fast new channel it cut only in 1965 between a sandy-clay wall to the north and the New Island to the south.

Here the river leaves my study area. It cuts through the high hills of the Jakey's Fork Moraine, that dam that once caused the Ice Age lake. There is a sort of canyon now between the moraine and the Badlands wall, a canyon I wish I knew more about. But it's too far for weekly visiting—I must turn here and go back.

A study of the creatures whose home is this mile or so of river ought to begin with the fish. I'm no fisherman, and I'm very poor at seeing them in the water. If it weren't for Ollie Scott, there'd be very little here about the fish. Ollie is a forest ranger who lived on our acreage while I was making this study, sharing the bottomland cottonwoods with us. He is interested both personally and professionally in the fish of our valley. Personally, he's one of the best fishermen I know, seldom coming back from an hour or two along the river without a big trout. Professionally, he works with the fish wardens in keeping track of the current status of this resource.

In late January 1967 he was taking part in a fish census of a 400-yard stretch of Wind River just upstream from my study area. The technique of these census takers is to introduce, by means of electrified rods, just enough electric charge into a short stretch of water to temporarily stun all the fish, so they will float up into sight and be counted. You work up the stream, so the stunned fish will slip on downstream and not be counted twice. You repeat the process many times, till the whole 400-yard stretch has been electrified and counted. Not all the fish get caught in the count. The charge is not enough to stun some of them; others swim through the count area too fast. But enough were counted to make my eyes pop. There were eighty trout, a smaller number of whitefish, and a few suckers. About 60% of the trout were rainbows, about 40% brown trout, and there were a few black-spotted natives, or cutthroats. From this count Ollie estimates about 1600 trout per mile, weighing a total of perhaps 500 pounds.

Ollie claims this stretch of river is phenomenal in its carrying capacity for fish. At present it is nowhere near its limit. The reason for this great capacity is fish food. Two warm tributaries come in a few miles above my study area. They are Big Warm Springs Creek and Little Warm Springs Creek, known always as "Big Warm" and "Little Warm." They keep the river from freezing and allow the algae to grow all winter long. This shelters and feeds all the forms of animal life that become fish food. In most trout waters, says Ollie, the fish stop growing in winter, just live along the same size until spring. But in this stretch they grow vigorously all through the year. At the time of the January census, the water temperature was 40°.

I asked if all the trout are planted, or do they spawn here? Ollie said, sure they spawn here, the rainbows in June, the browns in the fall. They migrate to spawn, but the migration may be for only a short distance, till the fish recognize what looks like a proper gravel bank. Mostly, the browns go up Jakey's Fork to spawn, but sometimes they lay their eggs on appropriate gravel banks in Wind River.

Whitefish run in schools. The river may be empty of them for a long way, then suddenly full. Trout, by contrast, are mostly loners, and their population is scattered quite evenly along the river. Each individual moves about a good deal. You can't count on a given trout being in a certain pool. Ollie hasn't caught a whitefish in six weeks, but trout he gets every fishing day.

Rainbows and browns feed in swift water, cutthroats back in the lower part of the pools, brookies in still water. There's always a drag of slower water along the edges of a fast stretch; Ollie has caught both trout and whitefish in water like that, next to the rapids near the Tripp's house.

Though it never freezes, Wind River never gets warm either. Forty degrees in midwinter, it's likely to be about 50° in midsummer. In the coldest weather the ice creeps out several feet from each bank and forms long arrows downstream from projecting boulders. Between them the green water flows slow and sibilant, loaded with ice crystals. Maybe the water is warmer here than it is a few miles upstream or downstream, where it freezes clear across, but to me the coldest sight in the world is a dipper taking a bath in this midwinter water. The dark little bird with turned-up tail stands on the pale ice below a cobblestone, right at the tip of the ice-finger where the green, sighing water sloshes over his feet and the icy fog of dawn wraps him. He splashes with energy, throwing about small drops that freeze white before they drop back into the river. I stop looking, for in spite of mittens, I can't hold the cold glasses any longer; but the dipper is still having fun.

Dippers are winter visitors in my stretch of river. This is their idea of going south for the winter. They wouldn't think of summering here; the river is far too calm. They need wild mountain torrents, preferably waterfalls, where the nests will be continually

wet with spray. I know a nest near Torrey Falls, ten miles south in the Wind River Range, in a narrow canyon, where the wild stream bellows and the dark rocks always glisten with spray. A pair of a dippers has been nesting here for many years, probably for many generations. I like to think these are the ones that winter with me, coming the last week in October and leaving around the middle of March.

These embodiments of water spirits aren't really water birds at all but relatives of wrens. They're a little bigger than most wrens, but hardly bigger than the big rock wren. They've no webs on their feet. The only adaptation to water life seems to be rather oily feathers—when they walk about under water they wear silver coats of air.

Walk about under water they surely do. Their diet looks to be snails, slugs, hellgrammites, and algae from the stream-bed rocks. Without any fanfare one will walk from a rock down into a pool till his head is covered. He will poke about the rocks for a while, then walk out or fly out, light on a rock, and shake like a dog. I have watched them repeatedly fly into, and sometimes right through, the big rolling wave at the riffle below Lower Island. It looks like fun.

In late winter the river gets very shallow—only knee-deep to a dipper in the riffle by Mabbott's house. Here is the spot where I've most often heard this versatile little one singing. The song varies. Sometimes he is warbling and chortling in separated phrases, a little like a warbling vireo. In February I watched and listened to one standing on a cobble in mid-river, shaking with the intensity of his full-throated song, like a mockingbird's, with chortles, warbles, and long swinging passages. Dramatically, he would at intervals raise his dark head and close his white eyelids. A week later I watched and listened to him singing from an ice-cake. Most often he has been standing under the bushes at the very edge of the water, almost hidden and with no airs whatever, just opening his mouth and letting the song drool out. Every once in a while he interrupts his song with a wren-like scolding note, a harsh gutteral chirring.

Because of the open water, our river is a winter haven for ducks, and the colder it is, the more ducks we have. Most common are mallards and Barrow's goldeneyes. There's a sprinkling of common goldeneyes, enough to keep me checking: if the white spot on the black face is a crescent, it's a Barrow's; if it's a circle, it's a common goldeneye; and Barrow's has much less white showing on breast and sides. Both mallards and goldeneyes are dabbling ducks, and where there is so much food in the shallows there is no unfriendliness, so you find both species together. But at the slightest alarm they take off by species—the goldeneyes whistling low, flying parallel to the water and only a couple of feet above it, the mallards quacking loudly as they climb steeply up above treetop level.

The mergansers are different. Common mergansers are around most of the winter, but not every day. They are more restless than mallards or goldeneyes, with more drive for exploring. Since they feed on fish, they require more hunting space, just like other predators; and like other predators, they seem to have more reasoning powers and more sense of fun and games.

Other ducks, less often seen, are pintails, gadwalls, the three teals (blue-winged, green-winged, and cinnamon), scaup, shovellers, baldpates, buffleheads, and red-breasted mergansers.

Wind River is a mountain stream. When you wade it in fishing boots, you find it's faster then you expected. When you are brave enough to swim in the few pools big enough, you find that (between the current and the water temperature) your best bet is to copy the mergansers—let yourself in at the top, splash down to the bottom, come out shaking and gasping, then soak up warm sunshine as you make your way ashore and back up to the top by way of the bank. Probably you'll try—once—fighting the current upstream. You'll get tired out and chilled before you make it. These factors of experience deep in my blood make me puzzled as I watch the gatherings of goldeneyes and mallards, feeding and grooming in the early winter sunshine, looking as calm as if on a barnyard pond. Yet that Wind River current is flowing past them and surrounding them. Why aren't they either flowing with it or giving some evidence of fighting it? The only answer must be that those webbed feet are paddling like crazy underneath the placid bodies. I am reminded of a group of old ladies gossiping on a sunny porch, bodies immobile while tongues, fingers, and knitting needles fly.

The first order of business on a winter day is grooming. Ducks always look neat, but it takes time and concentration to get and stay that way. One stands upright in the water, webbed feet curled around a slippery cobble, running his beak carefully over and over down his breast feathers. Then he stretches first one wing, then the other, at the same time arranging all the covert feathers with that broad blade of a beak. The final touch is to reach straight back, press the oil gland at the base of the tail, and distribute the oil where it will do the most good.

A duck concentration brings predators. Bald eagles, and once in a while goldens, hang around our stretch of river in winter. There are four trees they frequent, all snaggy cottonwoods with stout dead branches. These are the Eagle Trees. One is in the Lower Pasture, one in the Slough Woods, one near Tripp's house, and one near Leseberg's corrals. These are good spots for them to watch from. We do see them at other places, too, most often soaring in wide circles overhead. On a January 9 a bald eagle, swinging down the valley, saw me standing near the bridge. He descended to perhaps fifty feet above me and swung slowly in a half circle, keeping his head pointed to me. Half a circle must have been enough, because he went on down river, leaving me with the sharp memory of his sunlit white head and tail and the perfect pattern of the broad dark wings outspread.

Once Joe and I watched a golden eagle kill a mallard. That was a year when our neighbor, Gordy Shippen, raised oats in the field north of us. Enough good oats were left on the ground after harvest to bring for the winter the most mallards we've ever seen here, a couple of hundred anyway. That morning as we were eating breakfast a mass of ducks flew south past our window. We jumped up and reached the window in time to see a great golden eagle reach forward his talons and take a mallard right in mid-air. They were hardly a dozen feet above ground, and the weight pulled the eagle lower. He recovered before touching ground and flapped heavily across two barbed wire fences before touching down. Right at that spot he ate his breakfast, pulling the duck to pieces, looking around fiercely between bites.

On December 14, during the duck hunting season, I found a crippled mallard hiding near the bank by our water-gate where a stooping red birch made a dark shadow. One wing was shot, drooped helplessly, but the duck swam fast into the stream away from me. Later the same day, a hundred yards downstream on the Motel Bar, our neighbor Bob Tripp watched a bald eagle eating a mallard. The duck was not yet dead. Though its body was firmly held by the great toes and talons of the eagle's foot, one good wing still fought futilely, while the predator pulled out chunks of feathers and tossed them brusquely over his shoulder. A couple of magpies sat a few feet away, catching each bunch of feathers and dissecting it with care. My guess is this was the same crippled duck I'd seen that morning.

I had a dramatic sighting of a bald eagle one February day in '64, a day of low-hanging ribbons of fog. The big fellow was flying downriver under the fog, first coming straight toward me low over the trees, then veering off and rising, swinging toward the ragged red rocks of the Badlands, then circling above them, among swirling mists.

A comical contrast was an encounter, a year later, between a bald eagle and a magpie. Usually the magpie looks like a pretty big bird, but in this situation he looked absurdly small, but as brash and cocky as ever. On the same branch of the Leseberg Eagle Tree, he half lifted his wings and loudly told off the giant. For his part, the big boy thought it best to look impassive, shoulders hunched, hooded yellow eyes focussed far down the valley, great hooked beak clamped tight.

There was a puzzling sighting on April 5, 1966. I was down in the Lower Pasture when a big eagle passed, flying slowly upriver, low over the water, between the trees on each side of the channel and on out of sight. Suddenly from the direction of his disappearance a great blue heron shot up into the air, flew east as fast as it could go, much faster than typical heron flight, neck extended in an un-heron like manner. Faster behind came the eagle, gaining, and not more than a hundred feet away. They flew over the moraine hills towards Jakey's Fork, and I never saw what happened next. Did the eagle want the heron or just the fish in its throat?

Great horned owls take ducks sometimes. I just missed seeing it happen one day in early February, in the eerie light of dawn. It was the duck's loud squawk from Motel Island that made me look. A dark, heavily weighted bird was taking off across the channel and flying into the shadowy cottonwoods—a great horned owl with a mallard in his talons.

Predatory mammals don't seem to bother the ducks much—except for man, that predatory mammal that bothers everybody else, including ducks. All through the duck season the pickups wheel past, stop at the bridge, let out the hunters who trot up and down the banks. The banging of the shotguns becomes a background sound we try not to notice. The ducks' treatment of me makes me ashamed for my race. Surely they know me, seeing me day after day harmlessly strolling the dawn-lit riverbanks. I am always accompanied by at least one dog, and sometimes we are joined by a real retinue of all the neighbors' dogs. The ducks never fly up for them, however noisily the dogs may be skylarking along the banks and splashing in the water. It is I who am the enemy. When I appear they leave, the goldeneyes low and the mallards high.

Bobcats, they tell me, are quite common. They are supposed to eat mainly rabbits, and they are said to catch ducks when they are lucky. I don't know. My experience with our bobcats is limited to two occasions. Once I photographed tracks made by a big one crossing the bridge after a light snow. He was walking, and the space between his footsteps was eighteen inches. Maybe he was the same big fellow that I watched in lordly and contemptuous action another time. Joe and I were awakened one summer morning by our dog's frantic barking. Not twenty feet outside our window sat a bobcat on the gravel of the parking space. Ten feet or so away circled the dog, telling the cat (loudly) what he would do to him if he ever decided to use brute force. The cat held his striped face haughtily high, casting never a glance in the dog's direction. Something was bothering his furry right ear. He raised a hind foot and quite gracefully scratched it. He stood up, taller at the withers than the shepherd dog, and deliberately strolled past the side of the house, paying no mind to the dog, who was circling him at a proper distance and making plenty of noise about it, yelps rising to crescendo as the cat walked, subsiding to diminuendo as the cat sat down again, arching his back and looking through and past the dog at the fields, trees, and house. Fascinated, we ran from window to window to see all the show. The cat half-circled the house, letting us clearly see his heavy tan fur dappled with uncertain spotty stripes, his big fuzzy ears, his little short striped tail, the arrowy black markings about his face, his long legs and big feet, and mainly his sinewy pride. Three times he sat down, gently scratching his right ear each time. When he disappeared behind willow bushes in back of the house, the dog trotted thankfully to the front porch, obviously saying to himself, "Well I'm a fine watchdog. I ran off that big boy. And if he'd made one false move I'd have torn him limb from limb!"

Another predator along the river is the kingfisher, a raffish noisy extrovert who has no notion of concealing himself. As a predator, he bothers no one but the fish.

At sunrise one day in late November I watched from the bridge as one sat on his regular perch, a red birch branch over the gray-green water. Just after I passed, I heard him rattle, turned back just in time to see the splash as he went under water, and I didn't see him again. The next morning, though, he was back unhurt at the same stand. A few days later I saw him catch a fish from the same place. No preliminary sound this time, just a belly-whomping splash and a U-turn at the water, with a twist at the top of the U, bringing him back to the same birch branch with a trout maybe five inches long in his bill. A quick flip of his big ragged head, a long swallow; then he pulled in his neck, aimed his spear again at the water, shook his shoulders, and set his sharp eyes to watching.

On Thanksgiving Day I watched him catch five small fish one after the other from the same perch, about five feet above the water. He threw himself at the water, splashed, swung up to the branch, swallowed briefly, then down again immediately, five times. A happy Thanksgiving for him, less so for the fish.

Once Ollie caught a trout that a kingfisher had hit, but lost. It showed a deep wound behind the head, the stab from that wicked beak. It seemed otherwise healthy enough and of sound flesh, except that it was paralyzed on the side where the stab wound was.

Kingfishers are noisiest in pairs. They seem to love a chase, playing a loud game of tag upriver, down, and across, in three dimensions, diving, somersaulting, turning impossibly sharp corners. Only once have I seen this happen when it could reasonably be thought a courting flight: that was in April. Other times when I've seen these wild chases are September, October, January, and February. They may be males fighting over territory. I seldom see kingfishers in summer, and I don't think they nest in my studymile. All winter long one is here, and sometimes two, from September to April. He sleeps a little late, or they both do. There are holes in a vertical red clay bank a couple of hundred yards downriver from our "river-house." He comes from that direction a half-hour or so after sunrise, clapping his blue-gray wings in a syncopated rhythm, generally heading straight upriver to the red birch branch. One time I saw him strike right out across the greasewood and sage, past the edges of the Badlands and across the Painted Hills subdivision, his white belly dramatically lighted from the low sun. When I do see two, I can seldom see whether one has the brown belt of the female.

To me, the most exciting winter resident of the water realm is the moose. We are quite a long way from the moose's usual environment. The willow thickets of the high mountain meadows and the spruce bogs are from eight to thirty miles away, depending which road you take. But almost every winter I see moose, sometimes at much too close range. Maybe the ones I see in the Slough Woods or on the New Island have made their way downriver all the way from Sheridan Creek. Maybe they've come down Jakey's Fork from the Moon Lake country and then upriver from the mouth of Jakey's Fork.

Where they come from, no matter; they're always scary to meet. One March 15 in the early dawn I was startled to paralysis by a sudden loud crashing in the silverberry brush right beside me, followed immediately by loud cries from Buttons. Right away she put a big calf moose, probably a coming two-year-old, across the cobbles in front of me and into the middle of the river. He turned to face us both, pawing the water, head low, little eyes gleaming, tongue going in and out, black hair standing straight up on the back of his neck. Wow! In one frantic second I was able to think only that his mother must be close behind me, there weren't any trees nearby big enough to climb, and maybe this was the end. There were a few tense seconds while I stayed paralyzed. Then the calf swung around and trotted splashing past the end of Lower Island, up past Shaw's old corrals, and out of sight. I put the thicket behind me and shortened my walk. The rest of the winter I kept a sharp eye out for cow moose, but never saw one. She must have turned the big calf loose, in preparation for a new one about to arrive. Maybe she was near the mouth of Jakey's Fork or perhaps high up on Wood Hill.

Always in the river realm I watch for beaver. In my stretch of the river, there are no beaver-ponds, nor are there aspen, beaver's legendary favorite food. Still, the beaver are here. One fall they stashed a large supply of willow stems beside the rock-filled crib under the middle of the bridge. It's a good trick—I haven't figured it out yet: how do they make those willows stay put in fast-running water? They don't jab the bottoms in the mud. They don't weigh them down with rocks. A beaver stash is a neat pile of sticks, butts upstream and tips downstream, reddish stems side by side and many layers thick, seeming to waver a little as you watch them through the moving green water. Every day the owner of the stash would select his meal from it and carry it to his favorite perch on the crib. He'd then methodically peel the bark and chew it down, finally dropping the peeled white withe into the water to wander downstream as the current pushed and pulled. Before spring the river bottom downstream from the bridge looked like smocking, closely embroidered in a rick-rack pattern.

Not often do beaver try to dam Wind River and never in this stretch with success. Several years ago a pair worked hard all winter when the water was low, making a dam across the middle channel at Leseberg's Islands. They cut twenty-nine trees on the main island, incorporating their carcasses into a strong dam that rose higher and higher until it was almost above flood level. Though I mourned the passing of those lovely trees, still I warmed to the beaver's energy and creative ability as they strengthened the dam all through the wintery days and nights against the floods to come. And in vain; when the water came high and swift, debris pounded at the dam till it broke, and all their labor was nothing but a mass of plunging toothpicks soon disappearing, a cleanwashed pair of stream-banks, and twenty-nine weathering stumps.

With no ponds to build houses in, the beaver along here live in the banks. I know four beaver houses in my stretch of valley, none of them inhabited this winter. Last year, one of them had some repair work done on the roof, but the carpenters didn't stay.

The one that gives away the inside construction is the fourth, because it has been torn open by floods and neighbor boys. The house-builders seem to start digging in the edge of the bottom of the river bed, and make a tunnel more than a foot in diameter down, horizontal, then up at a slant. It would seem that they dig till they come out at the top of the bank, widening the top hole to two feet or so. Just below this hole they widen out a chamber to beaver-bedroom size, say three and a half feet across and fifteen inches high. Then they roof it with fine clean poles, cross-hatched and interlaced into a structure that is very stout. It takes determined boys and lots of effort to tear it apart. At intervals the carpenters fill in the cracks with a mortar of dirt and mud and

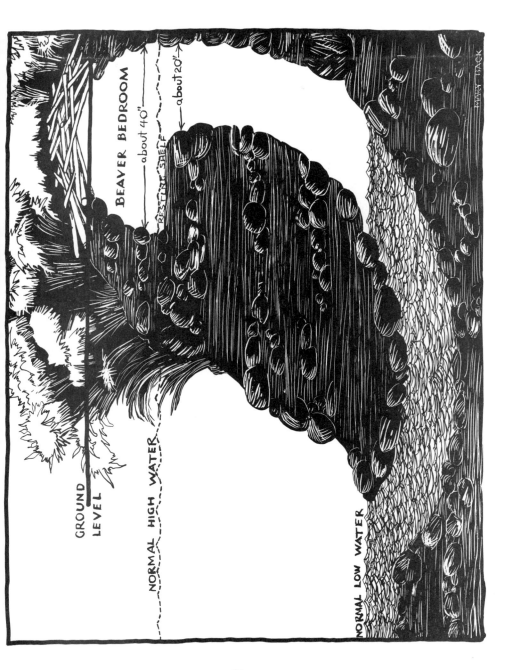

then work in another layer of sticks. When the house is finished, what you can see of it is a shallow conical pile of sticks and mud that looks as if a flood had casually deposited it, yet it is so strong that it will not quiver beneath the tread of a horse. In all four that I have been watching, the lower entrance is covered with water almost all the time; as one may say, "The river's so low the beavers' entrance is in sight." At the other end, the floor of the chamber is so high that it will take extremely high water to flood it.

The river is lowest in April and highest in June. I have a place to measure it. In the middle of the bridge, the log crib full of rocks that is the bridge's main support has an outer wall of ten-inch planks. Last June there was a time when I could see only one and a half planks above the snarling water, where in April I counted seven and a half. The difference of sixty inches changes the river's temper from pleasant murmuring quicksilver to doom-threatening pouring steel.

But spring comes to the river realm long before floodtime. You are aware of the low water that is a sign of early spring long before April. On February 7 the river was so low that at Lower Island I walked in the middle of the river bed, with the shrunken stream hardly twenty feet wide between me and the island. This is the time that the willows begin to color up. I do wish I knew more about willows! I wish I had a good reference on them. I read somewhere that the most rapid evolution of any life-form is taking place in the willows. Maybe that's the reason botanists hesitate to stick their necks out and name them—maybe the names will be outdated by next year. Anyway, the reference is likely to say *Salix sp.* and let it go at that. How many grow along my stretch of river I don't know. But this I do know—willows come in numerous fascinating colors. Their stems are not brown: they are raspberry, maroon, gray, blue-violet, amber, yellow-green. And whatever may be the natural color, the spring flood of sap intensifies it. Long before the leaves, long before the pussywillows, the sapborne color flash gives away the secret: spring is on the way.

Before that warm day, though, I have heard in the dark, competing with the whistle and whine of the west wind, the authentic herald of spring—the two-syllabled cry of the killdeer. Sometimes killdeer winter at the Big Spring of the New Slough, and we'll get to them after a while. But the wintering ones are almost silent, just whispering their call. This year, on February 26 I heard the two-note call, and I ran outside, shivering without a coat, and felt the great joy of spring in my heart, beating in time to that shrill minor keening in the darkness.

The next day I found a pair of killdeer on the narrow beach by the river-house and marvelled at their camouflage. Camouflage! What a word for such brilliant pattern: pure glossy white all over below, crossed on the neck with two black rings; dark head, pure white forehead, bright red eyelids, white ring back of head, olive brown back, and a gorgeous tail pattern—orange from halfway down the back, black across the fan, and white around the rim. But it still is camouflage. Against the round cobbles of the beach, some dark, some light, each bearing a crescent-shaped dark shadow, striped too with random patterns of chips and twigs and their shadows, the killdeer completely disappear. You see them only when they move, scampering along by the waterline, picking up a bite here and there. You fix your eyes on one, determined to be able to see him when he stops. Then he stops, and you've lost him.

The flocks of mallards and goldeneyes were still there, getting more restless as they feel spring in the air. They take off more readily, circle round nearby, and often splash down again close to the take-off point. Newcomers join them. On March 7 the mallard bunch had with it a grey visitor with a square white patch on his secondaries that flashes plainly when he flies. Gadwalls aren't common here, but a few times each year I see one in spring or fall.

On March 18 five newcomers replaced the mallards on the beach just above the bridge. The low sun outlined with brilliance five big Canada geese, heads high, watching me. Buttons had told me noisily that they were there, so I walked up carefully. Buttons was still barking down at them from the high bank, but they paid her no mind. They knew the enemy, and they were watching me. Shortly they took off all together, under the bridge, downstream into the east wind, rose and circled north, formed a wedge, and flew out of sight. These great, fine, capable birds aren't at all common on my stretch of river; more often I hear them overhead. Their deep cries make my neck hair prickle when they fly over, east toward Torrey Lake or northwest toward the Yellowstone.

March 21, the equinox day, brought a grey squall of rain and a new visitor, a tiny duck among the mallards. I watched it quite awhile before I was sure it was the green-winged teal. From the paintings in the bird books, I expected to see vividly that bright green streak over its eye and cheek. Every time I see this bird I have to teach myself again that the thing to remember is: it's the little duck with the big brown head. You have to be pretty close, the light must be good, and the duck must be quiet, to see that green streak on the face. And the green on the wing that gives it its name? Oh, just forget that, you can hardly ever see it.

At sunrise on Sunday morning, April 10, a flash of color across the river made me focus my glasses on three little ducks, sheltering close together under the bank near Bob Tripp's house. Two were plain little brown females, and the third provided the name for all three. The male cinnamon teal had as bright a breast as a robin—he wasn't so very much bigger, either. The three stayed together in almost the same place for a couple of weeks. They should be nesting not far off, but I've never seen cinnamon teals here in summer.

The blue-winged teal, as tiny as the others, is the only one to nest in my study-mile. I've never seen them except in breeding season, not as early in spring as this.

March 29 this year brought the blackbirds. Our commonest blackbird is the Brewer's, and from now on till October they will be neighbors of mine. They are not commonly birds of the river realm; mainly they're in fence rows. But my first spring sight of them is likely to be a ball of blackbirds rolling up the river—a big transparent sphere filled with a hundred or two hundred birds, each bird free to change his location at will within the sphere, while somehow the shape of the whole ball stays intact. I was at the house on the highway, fighting to pin clothes on the line in the teeth of a brisk west wind, when I saw this ball of blackbirds turn the downstream bend of the river and start rolling slowly up toward the bridge. I forgot the laundry and ran to meet them. The ball was about twenty feet in diameter, and was rolling right on the water level, on my side of the river. Chattering blackbirds caught their feet on the cobbles, ran a few steps, pecked at the ground, hoisted themselves into the air, circled around the ball while others made conversation! What dippings and risings and flirtings of wings! Each blackbird was using a lot of energy, and flying quite fast, in short swoops of twenty feet or so at a time, downriver, upriver, or across the ball. Yet it stayed a ball, and progressed upriver at a stately rate of maybe two miles an hour.

It rolled right through the bridge. For half a minute or so the log trusses were blackened with restless whistling birds. Some lighted on the planking, then quickly left. The ball seemed not to break up at all, it unformed and reformed so smoothly, and rolled smoothly upstream and out of sight.

These are the blackbirds with white eyes, the most common species in my field of study. There are redwings in the cattails; a few pairs of cowbirds; the thrill of an occasional yellow-headed. Grackles are just extending their range to here—a pair or more each year lately, very sassy. Bullock's oriole is uncommon. The western meadowlark sometimes nests in my study-mile, but not always. None of these are birds of the river. But spring, the river, and the balls of blackbirds come together in my mind. This ball may have in it all the Brewer's blackbirds for the Upper Valley. In a day or two the flock will disintegrate as the pairs separately choose their nesting sites.

The great blue heron seems tropical to me. His mere presence in our high mountain valley seems wrong. There is a dreamlike non sequitur in his being here at all. April 2nd was anything but tropical. The ground was white with new snow, there was no sunrise, the clouds hung long fringes down toward the ground, and the cold air was congealing into more snow. It was somehow shocking on this grey day to find a great blue heron hunched in the water near the river-house, grey as the fog, except for the cold yellow eye turned toward me, just about the level of mine. A long exchange of stares, and he decided he had enough of me. He spread great wings like sails, and was off downstream, staying in the tunnel between the trees, floored by the grey stream and roofed by the fringed grey fog. His long neck was extended at first, but as he reached cruising speed he deftly folded it up, till his crest lay down between his shoulder blades, and his long spear of a beak lay half its length upon the crook of his neck.

April at our altitude is the month most of us could do best without. Those who can choose their vacation time are likely to choose April. We all know that seventy miles down the valley it is early spring, that Arizona is ablaze with desert flowers, and that for us there is snow ahead. We remember with a touch of bitterness the rhyme we learned in school: "April showers bring May flowers". Our April showers will be white and will pile up.

The winters themselves are not extreme and are easy to take. Our part of the valley gets very little snow. The upthrust of the Buffalo Plateau (11,000 feet, at the head of the valley northwest) splits the storms. Day after day we watch them go from west to east across the rugged canyons and peaks of the Absarokas north of us, and northwest to southeast along the high-rising timbered foothills of the Wind River Range to the south, while over our valley the sky stays blue. We know that the snow will pile up twelve, fifteen, twenty feet deep where the highway crosses the Continental Divide at Togwotee Pass thirty miles west. We know, too, that the split storms will coalesce again to the east of us, so that both Lander and Riverton, down the valley, will get more snow than we.

And alas! We know, too, that April is coming, to make up for the open winter. The April showers making green the lawns of Lander will be piling white stuff on ours.

April 13 was one of those days: dark, no sunrise, fresh whiteness all over the ground, the river a gunmetal sheen, hissing as the snowflakes hit it. Alongside it the light-wires leading to the Red Rock Ranch Motel were thick white ropes—except for four black exclamation marks slanted across a wire! Incredible! Here were four violet-green swallows surveying the unlikely landscape. Precious little outlook for insects; and air-borne insects are all that swallows live on. These four are six whole weeks earlier than my first violet-greens of last year. I can't believe it! They let me come up for a closer look. They've dusted snow off the wires with their narrow dark wings and now sit quietly composed. No doubt about their identity, their field marks show clearly—the white clown faces, framed with neat purple-black skull caps, and the white spots on each side of the rumps.

What can they be living on? A short while later, I find they aren't so dumb. On the way home, I walk through a couple of swarms of midges, dancing up and down in the cold snowy air. I guess the swallows know. There are fifteen violet-greens skimming over the water a week later, twenty-one the day after that.

We've too great a variety of swallows here. I use up a lot of time watching their aerobatics, trying to follow each bird in turn to find out what kind he is, for each group is likely mixed. They are bobbins scooting through invisible warps, endlessly weaving invisible tapestries, and so fast that each bobbin is nearly invisible itself. The violet-greens and the tree swallows both nest near our river-house. They look much alike, being iridescent dark above and brilliant white below, but the white faces and rump-spots make the violet-green easy to tell from the other. Bank swallows and rough-wings are both brown. They both live in holes in river banks. You have to look closely to see the dark band across bank swallow's breast, where rough-wing's breast is white. Cliff swallows have apartments high in the twisted narrow canyons of the Badlands. They don't hunt as often over the river as the others; when they do, you can spot right away the pinky-rusty rump and forehead. Barn swallows are scarce, but every year I see a few. They seldom nest in my study-mile, else I'd see more of them. Perhaps a pair lives in Leseberg's barn, or maybe comes up the river from Fish's; one year a pair built in a dude cabin at Red Rock Ranch. The "swallow tail" and bright rusty throat are sure marks of identity.

Although this area is not a flyway, there are exceptions. At the head of the valley, at 9650 feet, is Togwotee (pronounced Two-Go-Tee) Pass, an easy pass for birds and people most of the year. The forest flows right through it, so there is good shelter in case of storm. Mostly the winds are not too strong or the cliffs too scary. So in season we sometimes see white pelicans enroute from the Gulf to Yellowstone Lake. Whistling swans are sometimes migrants, too. And the fantastically rare trumpeter swan is sometimes a joy to our eyes, for we are just over the hill from their breeding ground in Jackson Hole.

Once I saw a snowy egret, of all unlikely sights the most unlikely. On May 14, 1970, Bob Tripp phoned about noon for me to come look at the bar by his house. I watched that wader beside the bar for ten minutes at about 100 feet with eight-power binoculars—no doubt at all about his identity. He was a smallish white heron about twenty inches tall, with lovely white plumes on his head and breast and down his back. His bill was black, his cheeks were yellow, and it was astonishing to see that at the ends of his long black legs there were big yellow feet with black nails. The egret was working hard, walking slowly around the edges of the gravel bar in water about five inches deep. He was scuffling constantly with his feet, watching the moving water with sharp eyes. I saw him catch three fish in succession, three inches long or thereabouts. When he left, he circled slowly above the trees, then coasted as if to land near the bridge. I hoped to see him again, but I never did.

The gull and the osprey are not migrants. They live around here and I see them every once in a while. Or maybe I will never see an osprey again. There is a nest some generations old, a great ragged rack of sticks on top of a limber pine at the head of Torrey Lake, about four miles southeast of our studio. The pair using it had a regular beat, up and down all the streams hereabouts. Then some vandal shot the female as she stood on her nest. According to the literature, osprey mate for life. The next year the nest was untenanted; but one osprey, like the one this morning, was occasionally seen alone.

I don't know where the gulls nest; there are so many lakes that might be candidates, but none are within my study-mile. The California is our commonest representative of its family. This is the species immortalized on a monument in Salt Lake City, an expression of thanks by the Mormons to the gulls that came in miraculous flocks one plague-ridden summer and devoured the "Mormon Crickets" that menaced the crops. In my valley I have never seen a flock of gulls nor yet a plague of Mormon crickets. Perhaps there must be the one to bring the other. I have seen flocks at Jackson Lake, and screaming masses of them at Old Faithful Geyser in Yellowstone Park. On Jackson Lake it is probably the catches of the fishermen that bring them together. At

Old Faithful it is obviously the lunches of the tourists. Along our river, I have yet to see one alight, or hear one scream. They beat their way silently around the bends, generally at about the level of the tops of the cottonwoods, almost always just one at a time, his wings carving precisely lovely arcs against the sky.

May 25 I saw my first example of a different gull—Franklin's gull, wading around in the flooded meadow in front of the log cabins of the Red Rock Ranch. Much smaller and neater he was than the California gull. The glasses showed him beautifully groomed, with neat black head, light blue eyelids, blue-gray back, polished white breast with an undefined pink glow. When he flew, the black field marks backed with white on the wing tips were very clear. There were many Brewer's blackbirds on the same piece of wet meadow, maybe all after earthworms. I saw him only that once.

The river dominates this study area. Every spring brings changes. When the water rises I keep a close watch. On April 22, the river was up seven inches from the first of April. The snows in the mountains are beginning to melt. A high-and-low rhythm like the tides, is starting. High water at the bridge comes about 2 A.M., low water about 2 P.M. This daily rhythm will keep on till the snow cover in the forest is gone, sometime in July. As the water rises, everybody watches. Everybody hopes, and many pray, for an alternation of a spell of hot weather with a spell of cold, to retard the melt.

The river climbs the ladder of planks at the crib under the bridge. At first it murmurs, a deep rhythmic whisper. Then it talks loudly, arguing with itself. By June it is bellowing. It is frightening to hear and see; about a hundred feet wide now, filling its banks, mostly surfaced with long, brown, fast-changing scallops that look deep and carry narrow curls of froth on their lips. The earth vibrates, a fine tremor shakes the shore.

Debris is coming down—planks, poles, sawed timber from undercut piles at the mill, tree-trunks. We keep our eyes on the bridge, for we remember well the flood of 1950, when the river took the bridge out.

That was the first bridge ever at this point and the third one Joe had built over Wind River, the others at the two upstream ranches where the stream was narrower. It was made, like the present one, of logs, but unlike this one, it had no central rock-filled crib. The great logs used for floor stringers were set upon two heavy trusses, dividing the river into three streams. The whole structure was tied into a unit with plenty of iron straps and cables.

But a mile upstream the flood undercut the roots of four large cottonwoods and slowly tipped the big trunks till they fell into the water and moved with the current, roots first, high leafy crowns like sails behind.

That night Joe and I were sleeping lightly, conscious of the bellowing river, when we were wakened by a loud new sound like a sledge-hammer blow—then another. We pulled on shoes and jackets over pajamas and ran out into the sweet-scented, moon-silvered night. Hand-in-hand we ran toward the bridge. There was one more echoing hammer-blow, then only the cry of the water. We reached the top of the bridge-approach, but where the planking should have been in front of us there was nothing, only dark water racing by under the moon, a sheen on its surface, highlights on frothy ripples.

It was many months before we had another bridge. Nothing could be done while the water was high. When the water got low, it was winter, with frost and storms complicating the building. One of the great trusses had been carried a mile downstream, the other about 100 yards. Everything else was completely lost. Joe gradually assembled material—logs, cables, planking, spikes. With the help of a couple of neighbors, of a winch he had built from an old horsepower hay-baler, and of too much pure muscle-strain, he inched the nearer truss upstream until he could use it, sidewise as it lay, as the foundation for a solid crib that would hold many tons of rock, be a sturdy island in midstream, and hopefully make less likely a second such catastrophe. From the island to the shores there were now two bridges to build. All the agonizing slow routine, with only one man to help with the handling of the big logs, all the fighting of the storms, the ice and sleet. Well, it wasn't quite a nightmare, for there were solid satisfactions involved, but it was a struggle neither of us will ever forget.

We keep our watch. June 12, 2½ planks, June 13, 1½ planks. The river is beginning to leave its banks. Behind the motel, a loud-voiced new channel is racing through the underbrush. A few inches higher and the main river will be out of its banks near our river-house.

June 14, 2½ planks, a ten-inch drop since yesterday. The flood is over. Little by little, now, it will keep going down. It will be several days before I discover what caused the big sudden drop.

A half-mile east, downstream, the river made a big oxbow bend, curved nearly a quarter-mile southward, turned hard against the highway fill, then made its way north to the Badlands cliffs at the entrance to the canyon through the moraine. The bend surrounded on three sides a narrow peninsula covered with cottonwoods and willows. At high water on the night of June 13th, the river straightened its course by cutting across the base of the peninsula and heading from the east edge of Red Rock Ranch straight for the Badlands cliffs.

The first result was to relieve dramatically the upstream pressure, draining off excess water and ending the flood danger. Later results I shall be studying perhaps for years. The entire water table for the whole stretch of my study-mile has been lowered, in places as much as a foot. Erosion is faster, trout pools have been washed away, rapids are noisier. A big change is a wide expanse of barren gravel just below the Red Rock Ranch east fence where the river channel used to be. It's a new ecological life-zone, starting from scratch. It leads into another new zone, a quiet backwater shaped like a horseshoe half a mile around the curve. I named it the New Slough. Now everyone calls it Horseshoe Slough. The river creates as well as destroys.

Sloughs and Springs

THE BEAVER'S PROPER HOME is the marsh country. There isn't much of it in my area of study, but what little there is shelters some of the most interesting citizens of my world.

I have called the river the living water; maybe that sounds as if the sloughs and springs are "dead water." Not in any final sense they aren't—they are filled and pulsing with life. The only sense in which they are less "alive" than the river is their lack of muscular cutting power. The sloughs and springs lie quietly open to the sky, with only ripples to mark their current. They do not hustle and pressure their way; they encourage life to come to them.

There are two sloughs and three springs in my study area. The slough nearest the studio is just beyond the motel: I've named it the Motel Slough. It was once a river channel, but it has been a slough for more than the fifty years of my knowledge of this valley. Fifteen years ago it was a pond, with its water level held constant by a beaver dam at the east end. The dam was torn out soon after that by the rancher—the same man who shot all the herons—and the beaver took the hint and left. For years after, the level of the water fluctuated with the level of the river and was generally a narrow stream winding across a quaking bog. Somehow the beaver found out that the new rancher was hospitable, and they built a new dam, got a beautiful pond, repaired the roof of the nearby bank beaver house, laid in a big stash, and prepared to spend the winter.

Too bad for those best-laid plans! This time the trouble was the New Channel carved east of the study area by the flood of 1965. The lowered water table let the pond leak away out its bottom by midwinter; it's no good any more for a permanent

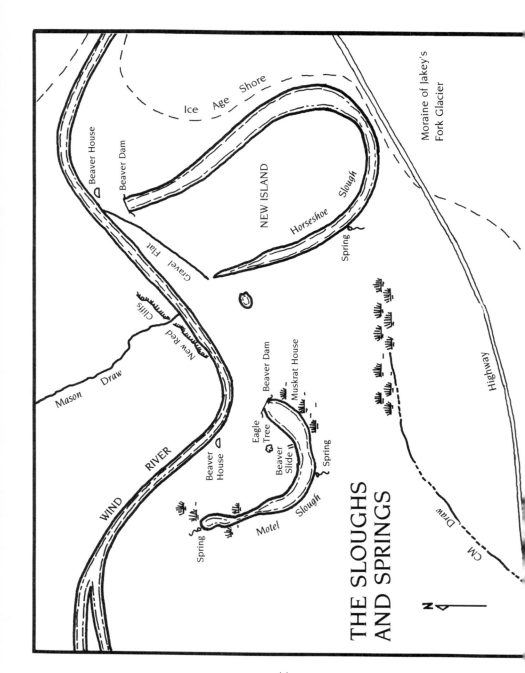

residence. But there's still some pond left, fed by two of the three springs. One makes a small pond at its west end and starts the little stream. I used to think it was underflow from the river, but since the water table was lowered I know better; it must come from some flow down from the valley side. It widens to a small, shallow pond, whose surface is generally covered with little green duckweeds; it narrows, then, to twist through a grassy swamp all bumpy with bog-knots. Halfway down, the stream's volume is doubled by a second spring, in the middle of the south side, away from the river. I think this one is the underflow from CM Draw, the dry wash that represents the drainage from the high benches leading down from Sheep Ridge and Wood Hill; it hardly ever carries surface water, but this spring is evidence that some water, out of sight, drains from those hillsides. It comes out in a steady pulsing that fills a basin underneath birches and silverberry, almost hidden by their spreading roots, a dark, mysterious, murmurous hollow, over whose outer lip the dark water, glinting silver, laps at the grass roots and carries its stream into the other one.

There's a cattail swamp beside the upper spring and another below the lower one. The pond between the second spring and the beaver dam near the river varies in size from wet grassy meadow to pleasant farm pond. From the first spring to the beaver dam, the Motel Slough is a horseshoe shape, walled with heavy underbrush, which now covers the new beaver dam. Beyond it, the final stretch to the river is a bar of black mud where I always look for the telltale footprints of the passersby.

Once in a while it will be a beaver, his wide webbed track as big as my hand. Usually it's the great blue heron in summer. His foot is even bigger. There's the spotted sandpiper, whose little three-toed foot makes a delicate lacy pattern. Sometimes it's the small, shockingly human, hand-and-foot track of the raccoon—shocking, because of the quick mental picture is presents of a human baby playing on that mud bar.

Beside the pond the beaver went to a lot of trouble to construct a playground. It looks to me as if the old ones made a "slippery-slide" for the young ones. It's a mound of mud all of five feet high, based on a willow bush, relying on the support of the willow's stems, helped out by some extra sticks woven in, basket-fashion.

Five feet high—that's a lot of mud to be carried in double-handfuls by animals weighing from forty to ninety pounds. Four slide-tracks ray out from the top, leading down into the water. When they play on this device the youngsters' wet bottoms keep the slides slick.

I am lucky to have two sloughs in my field of study. The one just described is old, with its place in the local life-design well accepted. The other is the Horseshoe Slough, the one made June 13, 1965, when the flood-powered river burst across the peninsula at the east end of my study-mile. When the flood went down, one could see that the main cutting power at that point was not eastward (in the downstream direction) but rather sideways, toward the north. A deep slice was cut into lake beds of many successive varves or layers of alternating red earth, sand, and gravel, overlying the present gravel bed of Wind River—a vertical red cliff whose raw newness is in places twelve feet tall. Surely someone who knew how to read them could learn from the varves a great deal about the age of the Pleistocene lake, could even count the years back and glean positive information about the weather of those years. I wish someone so prepared would read this and come help me.

The river now runs right against the base of this red cliff, ready to cut it away some more next flood time. South of the river is a wide gravel flat, representing both the old river bed and a hundred yards or so of lake deposits washed away that June night. This flat is seamed and wrinkled with braided channels, some of them carrying a little water that springs out of the gravel well away from its source, the river. The biggest of these channels is farthest to the south, right against the New Island that is the wooded remains of the old peninsula. It's a good jump wide at low water. There's one pond in the gravel bed with neither inlet nor outlet; since it is clearly down into the water table of the river, its water stays fresh, though well filled with long strings of algae, and by many other forms of pond life. In the years under my eye since the 1965 flood, I have seen plants begin to come in to the vacuum of the fresh barren gravel. Two-and-a-half years later, November 13, 1967, a quick listing included: willows, clover, slender wheat grass, tumbleweed, ragweed, curly dock, willow weed, aster, white top, foxtail, brome grass, gumweed, and cottonweed. Another development that is part of the ecology is the attraction the gravel flat has for a construction contractor from Dubois. Clean gravel for cement used to come from Mason Draw and across our log bridges. Now here is a new and excellent source, close to the highway and on its side of the river. So his diggers and loaders and trucks are often on hand, and here is another influence on the wildlife. The contractor probably regrets the plant life I am rejoicing in.

The Horseshoe Slough itself is shaped like a horseshoe open to the north and with its deep crescent against the highway embankment at the south. Like the Motel Slough, it is mainly watered by a big spring toward the west. In the winter when the water is low, there are several pools below the spring, filled with algae and watercress. Then the stream narrows and speeds up, twisting among some big rocks at the bottom of hard-frozen mud flats, and empties into a broad backwater that curves to the northwest and north against the Jakey's Fork moraine and finally all the way to Wind River. There is now a small beaver dam at its mouth. The outer length of the Horseshoe Slough, from the river south around the big bend and back to the river again, must be half a mile.

Off-hand, I would think that the birds of the river and the birds of the sloughs and springs would be the same ones, but it turns out that's not the whole truth.

Some of them, sure enough, are the same: mallards, for instance. The sloughs are their natural nesting grounds. There are quiet waters for fresh-hatched ducklings, lots of varied feed on shallow bottoms, a great variety of shelter. On June 3, I slid carefully around a bushy red birch, hoping for a glance at unscared bird life in the Motel Slough, but a pair of mallards immediately rose with a splash and a quacking and disappeared. There were other things to see from my vantage point, so I stayed quiet. A song sparrow watched me nervously, a house wren scolded sharply and fell silent, six redwings moved about among the cattails, two magpies flew across the water and lighted in the dead tips of a tall cottonwood, a sora rail stepped daintily around a bog-knot. It looked as if the swamp creatures had forgotten me. There was a swish of wings, and the female mallard slipped past the top of my bush and across the slough to a landing on the grass of its inner curve. She lifted her head high and made a strong, sharp sound, then a series of quackings. Without knowing the mallard language, I was sure she was alerting somebody. Sure enough, a yellow excitement in the midst of the grassy tussocks between the mallard and me became a little procession of new downy ducklings, talking softly, swimming in a line—not toward Mother—but, at an angle down the pond and in quite a straight line, allowing for the tussocks in the way. Mother gave them a final direction, halfspread her wings in time with her urgent quacking, then abruptly turned her back on them and ran in the opposite direction till she disappeared in dark underbrush, swaying clumsily from side to side but covering the ground pretty fast at that. It was only a few seconds till I understood her strategy: the aim was to get those little ones well away from me. If she talked to them from the bank, I'd likely look at *her* and give *them* more chance to get away than if she'd plopped out of the sky and landed among them and led them off (provided, that is, that I couldn't understand mallard talk). Beside the point where she landed was a dense thicket of birch brush. She had directed the ducklings to the grassy point at the far edge of this thicket. Then, clumsy as she was, she ran all the way round behind it, fast enough so I saw her roll out, swaying as she ran, just in time to welcome the youngsters ashore at the grassy point.

How on earth could she tell those new little ones so they understood?

I counted seven babies that time, but I must have missed some. On the 25th of June, twenty-two days later, at the same place, a female mallard led ten well grown young swimming away from me into hiding among the cattails. I thought it was probably the same family. Many times all summer and fall I saw about the same number together in the Motel Slough. They became tolerant of me, watched me with curiosity like mine. They groomed themselves, ate duckweed from the water surface, tipped up their tails with their heads out of sight as they groped on the bottom, learned to fly, circled the slough as a family group. There was a group of ten in the fall and into November, December, and January. I saw them in both river and slough, a close-knit bunch. After January they seemed to split up, maybe some had been shot, maybe that's a natural change in duck families. Maybe sometime I'll find out.

Other species of ducks are rare at my sloughs. The other common winter duck, Barrow's goldeneye, I've not seen at the Motel Slough, but four or five pairs sometimes sleep in the quiet backwater at the foot of the Horseshoe Slough. The teals—green-winged, blue-winged, and cinnamon—once in a while find the shelter attractive.

Killdeer, like mallards, are birds of both slough and river. Last winter twelve of them stayed all winter at the Horseshoe Slough. The water never froze at the spring, and part of the lower backwater was always open. On September 20 I saw one staggering as he walked near the spring. A close look through the glasses showed that his right foot was swollen badly at the heel joint just under his body feathers and seemed to cause him pain if he touched down his foot, hence his lurching gait. Arthritis? I wondered. On the 4th of October, on the mud flat below the spring, I watched for a long time one who seemed to have lost his right foot at the heel joint. I guessed it was the same bird. He seemed active and well groomed and certainly was accepted by his fellows, though he was quite conspicuous as he hopped while the others walked. He flew well, and handled his one-footed landings expertly. That was the last time I saw a crippled bird. The wonder stays with me: did he survive? Perhaps his foot wasn't off at all. Perhaps it was just drawn up into his breast feathers for protection, and I saw him later all well, but didn't recognize him.

In the fall and deep into the winter the killdeer were almost mute: none of the vigorous minor-keyed screaming of spring and summer. When they rose from the mudflats they often flew silently or with a whispery echo of a cry. In mid-December the cries strengthened and in January began to sound like spring. On January 18, with the temperature 8° and falling and most of the New Slough frozen, it seemed incredible to hear the spring-like cries, while from fifteen feet away I watched eight killdeer wade knee-deep at the edges of the ice, with ice-drops weighting the tips of the breast feathers.

January 21 I had my first glimpse of one back on the river. It was even colder, 4° that morning. I was standing on the shore by the river-house when a killdeer flew upstream out of the icy fog, made a quick U-turn by Mabbott's water-gate, and passed me again, going downriver, crying like spring. That foray must have convinced him the river wasn't yet fit for living; I didn't see another one away from the slough until March 13. After that, it was really spring, and the killdeer again patrolled the islands and bars. On April 13, a pair mated just across from the river-house.

All the swallows, of course, weave their invisible tapestries over the sloughs as well as the river. During the short swallow season, I am forever frustrated, trying to disentangle the species and count the numbers of those flying bobbins.

Hawks aren't slough birds, but many of their natural prey like the sloughs, so hawks of many species follow them there. One of the Eagle Trees is close to the Motel Slough outlet. A tall dead cottonwood with a stout limb near the top, it's used as a lookout by bald and golden eagles and goshawks, as well as by the real water bird, the gangly-graceful great blue heron.

One November day I saw a goshawk there, smooth blue-gray back turned to me. I started toward him, stumbling clumsily on the frozen tussocks, and he turned his head to enjoy the sight. A complication appeared when two magpies lit in the same tree. The hawk opened his beak, looked annoyed. Four more magpies, flying upriver, turned from their course and lit in the same tree. All six began squawking at the hawk. He raised his neck feathers, greatly changing his looks, and screamed "Kee! Kee!" sort of *sotto voce* (most of the sound may have been too high pitched for me to hear). Finally he dropped off the perch, displaying lighter breast and striped tail as he flew very fast downriver.

I wonder that Cooper's hawks ever win. One September day I saw one chasing a small flycatcher in and out and around the dead branches of a tall cottonwood—unsuccessfully that time, as far as I could see. I suppose it was the same one four days later that almost got a kingfisher, attacking him in the air over the slough. The kingfisher took evasive action, with sharp short flights and rapid changes of direction. He got away, too. Two years later, I saw one in the same area several times, a smaller one than the first. I thought him a sharp-shinned at first, then realized he was bigger than a flicker, though smaller than a crow, and (after referring to Peterson's *Field Guide*) spotted the rounded end of his tail.

The first time I saw him, he flew hard and straight into a cottonwood tree whose leaves were sheltering sparrows and warblers. Five birds sprayed out. I think they all escaped, for the hawk flew to a dead cottonwood branch near me and gave me a malevolent glare. The next day he nearly attacked me: flew at me, circled me fluttering about ten feet off, put out his feet as if to clutch me, then pulled them in again. I got a good look at his barred, rounded tail, rusty striped breast, rounded wings. Two days later he swept from the slough down Red Rock pasture, empty-clawed, behind ten screaming killdeer. A week after that, I saw him rush a group of six magpies, who exploded in all directions. The magpies and the hawk were about the same size in body, but because of the long tails the magpies looked much bigger; no doubt at all, though, they knew who was boss. The next day, he flew right past my head in the Slough Woods and lit on a cottonwood branch hardly twenty feet away and not much higher than my head, and stooped for a long mutual inspection. That was the last I saw of him. The ferocious energy he displayed each time I saw him must surely have been nourished. He looked in the best of health. But in five encounters I'd witnessed he missed his aim. There must have been some unseen successes.

Just once, in midwinter, I saw a northern harrier over the slough, flying low circles, only a few feet above the brown tussocks.

There are other encounters, more typical of slough life. I'd like to see more of the little sora rail. It's supposed to be common in the right places around here, but I've known only one at the slough. I saw him, or her, eight times in the summer of '63, from the first of May to the first of September. Since the times stretched clear across the breeding season, I suspected a pair and a nest but never saw more than one at a time. That summer I first heard the surprisingly loud, grunting call. The picture that sticks in my mind is the August morning I saw the slim little fellow picking his way through shining water, head low, feet lifted high at each step, under arching dead willow branches, where the big south spring boils out under the bank—dark foreground, dark bird, bright background.

Early one Easter Sunday morning I crossed the Motel Slough channel at the mud bar at its mouth and started squishing up the wet meadow with the sun at my back. One of the tussocks exploded in front of me. With a sharp beat of wings a little bird left it, whizzed past my face, across the mud bar and across the river, and I watched him land in the grasses just above the beach. He was a ridiculous little mite, a feathered tennis ball with a strong beak as long as the rest of him, and little short legs gathered up underneath. I thought "woodcock!," then recognized my first common snipe. A woodcock is darker, his eyes are more wierdly set at the top of his head, his head stripes are crosswise, not lengthwise like this little fellow's. After that Easter morning I jumped snipe every once in a while in spring and fall. I've learned to know the low grunty buzz, their call as they zip off.

Some of the sightings are of extraordinary beauty. On a September morning, with the sky and water bright cerulean and the swamp grass brown and orange a snipe was knee-deep in shallow water, quietly sunning, moving only enough to shatter and lengthen the shimmery reflections of himself. A year later, a gray and overcast September morning, I had a closeup for several minutes at the Big Spring where first I'd watched the sora. The snipe was crouched upon willow roots reflected in dark water, looking me over. At last he burst off his perch, projected himself into the air, and was off up the slough. As he flew past my head he was drawing up the greenish legs that are his landing gear. On a bright morning in mid-May, I saw two shore birds on a small patch of young grass surrounded by water in the middle of the slough. When I focused the glasses on them, I was astonished to see they were a killdeer and a snipe, taking the sun side by side, and watching me. I walked slowly toward them, till I was on the very bank of the slough, hardly twenty feet away. They watched me carefully and quietly, with no nervous twitchings. After some time, the snipe made the ultimate gesture of confidence—or contempt. He laid his long bill right down the middle of his back, raised one foot up into his breast feathers, closed his eyes, and to all appearances went sound asleep.

Other shore birds use the sloughs as resting spots during migration. A couple of times I've seen the solitary sandpiper and the lesser yellowlegs. The solitary was there once on July 2, which date implies nesting; but that was just one time I saw him other than migration times. The spotted sandpiper, our common breeding shore bird, I've seen only rarely at either slough. He must prefer the faster water.

Most interesting to watch was a member of the shore birds a little bigger than the spotted sandpiper, seen at the Horseshoe Slough on June 21. He was the size of a lesser yellowlegs, and that's what I called him at first sight. A closer look told me different, and he tolerated a really close look. His back was similar to a yellowlegs', sort of tweedy black-and-white, but his most conspicuous mark was very much his own—a wide dark line through his eye, curving and widening down his neck, getting reddish toward his body. His legs were dark gray, not yellow. This had to be Wilson's phalarope. He was feeding in a very special way. With jerky nervous movements, head held high, he walked rapidly around in a small area of very shallow water, seeming to roil it up; then he would feed just as rapidly—quick dabs into the muddy water, quick swallowing, quick turn to a new spot; then a quick short walk to a new shallow pool, head bobbing at each step; and a repeat performance of first roiling the water, then feeding.

I saw a dowitcher just once in the slough. That oddly named shore bird looks something like a common snipe, and, by George, there was a snipe standing right beside him on the bumpy wet meadow at the bottom of the Motel Slough, both beautifully posed for handy reference. Just a little bigger, dowitcher was much more gorgeous, with a rusty-red breast, black bill, and white and black tail. Snipe was brownish and dimly striped. They stayed long eough for a good look, then off they went together.

Deer tracks in the mud often show they've been at both sloughs, but I'd never seen them there till one memorable July morning. A good early start that day brought me to the moraine high above the New Slough just after sunrise. A doe and her spotted fawn were standing heads down in the water near the beaver dam at the lower end. As they drank, the busy reflections round their feet made and unmade their picture.

SLOUGHS AND SPRINGS

There's a muskrat house at the east end of the Motel Slough, close to the cattails where I count on seeing red-winged blackbirds from March to September. Sometimes I see muskrats swimming in the channels near the Big Spring. On a July day one swam straight toward me down the main channel, among tall grass and long thick strings of wavering green algae. About ten feet from me he dove, continued to swim toward me underwater till no more than six feet away, then turned sharp to his left and disappeared for good, right into the bank. Half a year later, on a gray January morning when the thermometer stuck around zero, I saw one swimming in a patch of ice-rimmed water in the Horseshoe Slough. He climbed out on a rock and did a piece of careful grooming, as ice formed visibly on tufts of hair—smoothing down his fur with his hands, pushing back his whiskers, and finally passing the whole length of his tail through his hands, before he slid all that combed fur down into the water again, and swam away.

SEVEN HALF-MILES FROM HOME

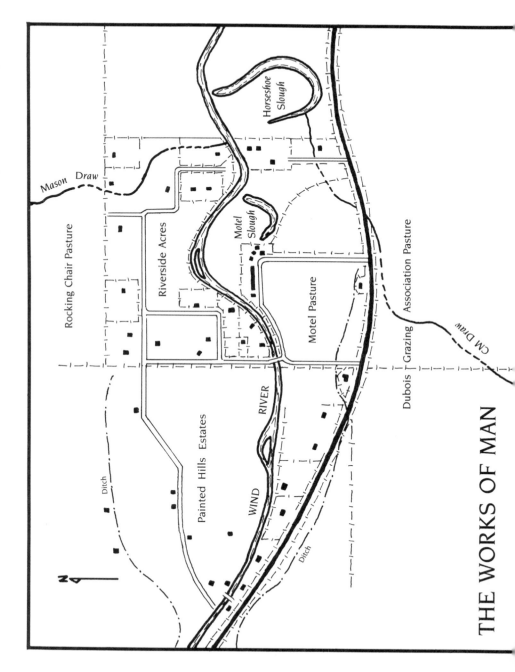

Fences

WHEN I STARTED THIS STUDY in 1963, three-fourths of my area was concerned with stock-raising. Two whole habitat areas or life zones were part of this concern: the hay meadows and the ditches. The studies I made as a matter of history and change still have value, but no longer as current ecological studies.

The Locke Ditch to the north and the Medill Ditch to the south are now historical artifacts. They have been abandoned and partly filled. All water now comes from the sky and from wells. The Leseberg Meadow that filled the whole northwest quadrant of my area is now the Painted Hills development. Enough houses are already on it to move it in my outline right into the "Works of Man, or Houses, Yards and Gardens." The same fate has come to much of the northeast quadrant, once the Red Rock Ranch horse pasture, now Riverside Acres, pretty well filled with thirteen homes. Happenings in this stretch are now told in "Houses, Yards and Gardens." Changes have come also to the Red Rock meadow south of the river. It is irrigated now by sprays from pipes filled with water pumped from the river right beside the meadow. The water is now used, not for raising hay to be cut, but for greening up horse pasture to be harvested by the teeth of horses.

With all the changes, the land still shelters wildlife. And though the use of fences has changed somewhat, the fences remain and have their chapter. They represent both obstacles and opportunities for wildlife and watchers. They keep horses in and likewise out. They limit the road area. They make places for plants to take root, thus attracting birds. And they are constant physical challenges to a bird watcher.

If you limit yourself to beaten paths, you will miss lots of fine places to see birds. But if you don't limit yourself to beaten paths, you are inevitably going to have to cross fences. So I have become quite a fence-crawler, though less than a skillful one, as snags on my coats bear witness. Joe bought me a new coat and remarked as he gave it to me, "Now, that is not for crawling fences!" Once I heard one Dubois housewife say to another in the post office, "Have you ever done Mary Back's obstacle course?" She meant fences mainly, I think, though I guess ditches, bogs, plowed fields, boulders, and riverside thickets also qualify as obstacles.

There are various kinds of fences, but in my particular study-mile they are nearly all of one kind, barbed wire, so that is my specialty. There are several ways of crawling barbed wire fences, and I think I have tried them all. Most used, and most frowned on by ranch owners, is the "bring a pal and stomp the wires" method. The pal puts a foot on the second wire from the bottom and a hand on the third wire, pushes with his foot and pulls with his hand and widens the space where you go through. Then you do the same for him. Do this a few times and there's a permanent sag in the fence. Then after a while a colt or a calf crawls through. Maybe he gets wire-cut in the process. Maybe he just gets on the highway and a car kills him. Whatever happens to the youngster, you've made fence-repair work for the rancher, who was kind enough to let you use his land.

Some places the fence is low enough, or built among boulders high enough, so you can step over, taking the chance of getting stabbed in the crotch.

Some places the fence straddles ditches or small gullies in such a way that you can squat down with a foot extended like a Cossack dancer, bend low your head and shoulders, and inch under.

But the best way of all is to lie down full length beside the bottom wire, then roll under face up. You can see and avoid all the barbs, you don't hurt the fence a bit, and it's good for your soul. It's a maturing thing to learn that you can willingly assume this humblest of all positions, then make a quick transition and swifly reassume the erect posture. (A warning! It's necessary to employ observation and common sense in choosing your crossing spot, else you end up with manure in your hair, rocks in your ribs, and cactus in your behind!).

FENCES

In this connection, I'll let you in on how buffalo do their fence-crawling. A few miles up the valley a rancher bought a few buffalo from Yellowstone National Park. (Anyone can; you can. Park rangers have a buffalo roundup every fall and the Park sells the surplus stock—surplus because after the calves of the year grow up, there are too many buffalo for their winter range.) The buffalo liked his place and did well. In a few years he had about fifty. But he could not hold them with any fence he had, and in the end this forced him to sell off his whole herd. A neighbor described to me how the buffalo handled fences: "A bunch walked up to this fence and a big bull in the lead looked it over. Then he went down on his knees. He hooked his horns under the bottom wire, and then just stood up. Wires twanged, half a dozen posts came right out of the ground, staples shot into the air in all directions. The old bull shook the wires off his horns, laid the fence on the ground, and the whole bunch just walked over."

My neighbors share with all ranchers the problems of fences. Fishermen and bird watchers may spread the wires, hunters leave gates open (or even cut wires so jeeps can go through) horses and moose break them down, and every year just weather, age, and dry rot break down a few posts. Mostly nothing can be done but mend the fence.

Surely anyone who treats his neighbors' fences with lack of consideration should have the educational experience of setting a few fence posts. If you haven't done it you simply haven't the mental background to imagine it. A post is heavy. It's seven feet long and six inches or more thick and weighs thirty to sixty pounds. The tools you need are a shovel, a heavy steel bar pointed at one end and flat at the other, a hammer, and a pocketful of staples. Also a wire cutter, a pair of pliers, and a roll of wire for splicing.

If your digging has been flower beds around the house and snow on the sidewalks, you are astonished at the really enormous labor of digging a nine-inch hole two feet straight down through hard clayey earth well sprinkled with rocks. You use the bar more than you do the shovel, prying the rocks loose. It's such an epic job you think the universe must be watching you. You finally achieve the full two feet, scrape all the loose dirt out with your hands, while you get dizzy from holding your head lower than your feet. You set the post in, shovel in a bit of dirt, then tamp it solid with many thumps of the flat end of the bar; then a bit more dirt and a lot more tamping, till at last the hole is filled and the post won't wiggle.

If you don't have to mend wires, then hammering the staples in to hold the four wires in place is a final triumphant flourish. But probably there are broken wires. You'll have to make loops in the two broken ends by bending the wire back and twisting the end around the shaft; barbed wire is astonishingly stiff and unyielding. Then you have to thread a length of splicing through both loops a couple of times and tighten it—and believe me, at this point you have to cave in and go find a *real* fence fixer: the beginner just has to be shown. When it's done, your face is red with the fine flush of achievement. How sad to realize that this is just the first step in "fixin" fence.

FENCES

Fences have their characteristic plants, mammals, and birds; and the fences most in need of fixing are likely the most interesting. Tumbleweeds or Russian thistles, bouncing down the wind, big as washtubs but almost as airy as balloons, catch against the wires by the hundreds. Their curly tan latticework catches the scanty horizontal snow that comes screaming down the west wind. The thistles hold it there till it melts; so the soil below it is a little wetter and better off than the desert land around. Seeds take root in the moist ground. Some were blown there, and, like the snow, were stopped by the thistles. Some are carried by animals or released from the droppings of birds. The hardiest live to grow up—the ones we false-proud humans call weeds.

Foxtail is abundant in the weed patch and is now becoming common in our open space. It is a beautiful and dangerous plant. Once in a while tourists or dudes bring me specimens to identify. They want the lovely sprays for winter bouquets and are shocked when I tell their gruesome story. Foxtail is a grass with a thick, long bearded fruiting head that really does look like a fox's tail except for the color, a warm light tan when ripe, not reddish. The long whiskers are delicately barbed, which makes them stick to the throat, mouth, and esophagus of the creature that eats it. Olaus Murie explained that the fine barbs harbor a fungus and hold that fungus in contact with the eater's mucus membranes long enough for it to start growing. Sores and ulcers form, which may become very serious. We had a Guernsey milk cow once who developed a foxtail ulcer that ate right through her cheek, so that pus dripped steadily down the outside of her cheek for a long time before it healed.

We had a husky pup who ate foxtail. Once she was coughing so painfully that Joe took her seventy-five miles to the nearest vet that night. He took a handful of foxtail out of her throat and charged $28.00. Joe brought her back, turned her loose in the yard, and saw her go straight for a foxtail plant and start to eat it. When we asked a vet about it, he told us that dogs occasionally become addicted to foxtail. There must be something pleasurable about those rough barbs going down the throat—something like whiskey, maybe?

Humans can suffer foxtail infection, too. Butch Wilson was the son of our Episcopal minister, and a star on the university football team. One year he had to have a throat operation just before school opened, which effectively kept him off the team much of that season. The doctor said the damage was caused by a fungus infection, probably from hay; Butch had been working as a hay hand that summer. It was guessed that he absent-mindedly chewed foxtail as he worked.

Once Joe and I were hunting deer high on the Torrey Creek ridge a few miles south of the study area. We were climbing steeply through a scanty forest of such junipers and limber pines as could hold to the slope. In a tiny flat-bottomed pocket about as big as our bedroom, with a view dropping awesomely down from its edge, we found the skull and horns of a fine bull elk. Our first thought was that a local hunter (who wanted the meat, not the trophy) had killed it there last winter; but a second look showed that the whole skeleton was there, almost all in place, the bones dragged about only a little by coyotes. We saw something queer about the jawbones and took them home with us.

I still have those two pieces. When we examined them, we saw that halfway along each side was a large lump of soft, unhealthy looking bone, roughly the size and shape of half an egg. After a little handling these lumps broke away from the solid bone, revealing enormous abscesses two inches or more long, eaten out beside two molars almost completely through the jaw.

When next I saw Olaus Murie I showed them to him. He said he thought they were the result of foxtail infection, as likely as not from grass on the Trail Lake or Whiskey Basin meadows.

I can picture vividly that grand monarch of the herd and can suffer with him his terrible toothache. When the pain grew so bad he could no longer eat, he deliberately picked that wideview pocket to die in. He just lay there and waited, neck extended along the ground, eyes fixed to the last on the canyon of Torrey Creek, and the Wind River Valley and Absaroka Mountains beyond.

Enough of foxtail. There are other fence-row plants. Knapweed and white-top send down their roots. Here come blue-eyed stickseed, red-and-green-striped curly dock, wide-eyed golden gumweed with its sticky flowers, feathery-leafed ragweed, wild sunflower, the giant dandelion called goatsbeard. In the tangle of annuals the shrubs begin to show up: the slender supple stems of sandbar willows, the spiny shafts of greasewood and gooseberry, the harsh, tenacious sagebrush, the stiff-stemmed, dome-shaped rabbitbrush. Here and there wild roses will make a massive hedge of thorn, actually hiding the fence. Near the river, where the moisture underfoot is more nearly permanent, the gray wild olive we call silverberry and the lacy red birch get a start.

Every once in a while, some big need for fence-fixing will cause one or both of the landowners to clear away all this hard-won foothold of the plants. New posts go in, fresh wire, hard-pounded new staples (but the old things are reused as long as possible; posts are turned upside down, the rotten part that broke off in the ground now high in the air; old wire is twisted on to new; all the old staples one can find lying around are salvaged). There's only bare ground under the wires and between the posts. But then here come the tumbleweeds bouncing down the field, fifty feet to a jump. They catch on wires and here we go again, as the shivering earth tries to cover her brick-red flesh.

From each such beginning, and all along the way, mammals and birds are attracted to the developing thicket, and they likewise help it develop. Little runways of meadow mice and white-footed mice are patted down among the stems. They bring along these paths their gleanings of grass seeds, and inevitably they drop some of these and cover them with the dust of their running, and some develop. As I come near the fence-row, the companionable and curious chipmunks scramble up fence posts and sit on top looking me and Buttons over, their fur stirring in the wind and outlined with a pale halo. Sometimes there will be three or four in a half mile.

Once I was startled to find a porcupine here. At least two miles from the nearest evergreen forest (high on Windy Peak to the south), the shelter of the fence-row must be most important to him, even though it's not much wider than he is. This is Buttons' first porky, so I hold tight to her collar as she whines, twists, and wags. The porcupine, big as a football, has his nose beside a post, all his quills erect, his fat black-and-white-spined tail lashing back and forth. I've used pliers to pull too many quills out of dogs' faces to want to take a chance with Buttons. It's hard muscular exercise for the human, great pain for the dog. So I pull Buttons well away from the lashing tail and hold her close to me for the rest of the walk. We never saw Porky again. Did he waddle, grumbling and rustling his quills, all the way up those two miles and thousand vertical feet, over the dry grass stubble, to the distant timber?

Cottontails find both food and shelter here. They play hide-and-seek with Buttons, who goes nearly crazy with excitement. She has never caught a rabbit in her life, and now has less chance than ever, with cataracts frosting her aging eyes. But that makes no difference to her fun, and the rabbits think it no worse than good exercise.

I've found the tracks of bobcats and raccoons following the fence. I suppose the mice and rabbits are the attraction. On a February morning, when an inch of fresh snow covered everything so that every mark had to be absolutely fresh, I saw clear prints of the humanlike front and hind feet of a 'coon crossing the pipes of the cattleguard where our entrance road goes through our line fence. Strange. It's a difficult place to cross. The pipes are parallel, six inches apart, and with a drop into a foot of space below them. And a 'coon could walk under the fence anywhere. Why did he cross the hardest way. Was it just the challenge? The fact that it was *there*, like the Grand Teton, and he hadn't crossed it before?

Deer and moose are once in a while in evidence along the fence. They aren't attracted to it as are these others. To them it's an obstacle, but not a serious one. Each has his own method of dealing with the problem. Does and fawns fold up their legs to a streamline with their bodies and jump through between the wires; or sometimes they crawl through as slowly and carefully as I do. Bucks jump over with a light spring; I suppose their horns prevent them from climbing through. Moose are amusingly languid about it. They walk up to a fence, stop completely, rise on hind legs, fold up front legs, and push their bodies over. Moose are not very careful about it. They are even worse than people about breaking fences, and even poorer as fence-fixers.

No threat to the fences, but just barely possible to find here, are our rare reptiles and amphibians. The only reptiles I've seen in forty-eight years in this valley are a couple of garter snakes and one water snake; the water snake was in Little Warm Springs Creek, the two garter snakes in a fence row. The only amphibians I've seen, aside from the tree frogs in the sloughs, were a couple of toads in a fence row, years apart. How can they possibly survive with so scanty a population?

Birds and fences really go together. Even a bare new fence attracts many, as perching spots and lookout posts. The more plants it acquires, the more species are attracted and the more members of each species. The balance of use shifts from lookout to shelter.

There are hawk species for each season: the fence-post hawks are kestrels in summer, roughleg in winter, red-tail in spring and fall. The red-tails usually nest higher up the valley, and they usually winter lower down. The roughlegs nest in the Arctic north, and some of them winter here. Both of these are big, beautiful soaring hawks whose main diet is mice. They spot the little fellows as they ride the air currents in big slow circles. They drop

to pick them up, then, if a fence is handy, will carry their tidbit to a post to eat it. They use the posts, too, for grooming, and for getting a rest from space. Kestrels have a more serious use of posts. They are responsible breadwinners, raising families nearby. The posts are real lookouts from which to locate the next food supply. Pretty little falcons of small-scale ferocity, they generally use food sizes in proportion—mice, small birds, grasshoppers. Grasshoppers must be a special rare delicacy. They eat a lot of them in season, but the season is short, from about August 1 till the first killing frost, which may be as early as the last of August, must be no later than early October. The rest of the year mice are the staple; the occasional bird taken is just a bonus. The fight with a magpie mentioned earlier was something else. The big black-and-white jaybird must surely have stolen eggs or otherwise disturbed kestrel's home.

Pheasants and chukars are two kinds of wild chickens we might see in the fence rows. This is the upper limit of pheasant range. One or two thousand feet lower they are abundant, even though heavily hunted. The shortness of grasshopper season may be a factor. Surely the winter climate wouldn't discourage them, for winters are always worse in Michigan, Wisconsin, Minnesota, and the Dakotas than here by Wind River; yet pheasants thrive there. Whatever the cause, the big handsome chickens are seldom seen.

Chukars are increasing slowly. Like pheasants, they have been imported from far away. Their native homes are the high deserts of Asia, such as the foothills of the Himalayas; and the high desert between the Wind River Mountains and the Absarokas seems to be just their cup of tea. They like to sun themselves on rocky ledges in the badlands, but too strong winds and bitter cold sometimes make the scanty shelter of a fence row seem worthwhile. Sometimes a scampering covey of upright little runners will head for a fence row's shelter or, just as often, will leave it for the open sagebrush as I approach.

One day, as we crossed the paved highway, Buttons saw a flicker of movement by the fence and made for it. Often she seems almost blind, but she does notice unexpected things. This thing she looked at closely for a moment, then picked it up gently in her mouth and brought it to me. It was a dying chukar, I guessed hit by a car. Hopeful that it might live, I carried it home in my hands, but it died within the hour. Such a beautiful bird, the only one of its kind I've ever held. Related to quail and partridge, it was bigger than a bob-white and smaller than a ruffed grouse, sandy colored, darker above and lighter below, with a bluish sandy band across its breast. Bill and soft chickeny feet were orange. The black accents were brilliant contrast. A parenthesis-shaped heavy black outline ran from its beak through its eye, in back of its cheek, then down to a point in mid-breast. When you looked straight from the front, the two parentheses made a compressed heart outline, built up of tiny, glossy black feathers. I got right to work making a water-color drawing of it.

When I see mourning doves they are nearly always along the fences, usually in groups of from four to ten. They sit on the wires, intoning happily that lonesome wail, or strut about below them, comically like stones learning how to use fleshy pink feet. Most of their food they seem to find on the ground near the fence.

Lewis's woodpecker I've learned to know along the fences, and I've seldom seen him elsewhere. He's the most stunning woodpecker you ever saw, with all black back and wings, red face, and bright shocking-pink breast. In habit he's as much unlike other woodpeckers as he is in color scheme. He acts like a flycatcher, spinning off from a fence-post perch after an insect, returning to gobble him down and to wait quietly for the next. Powerlines follow fence lines in several places, and these offer Lewis's woodpecker the tall poles he needs for nesting and for drumming, for when it comes to home life he follows the conventions of his tribe.

Most of the flycatchers are fence post birds, kingbirds most often. Olive-sided flycatchers and wood pewees are sometimes there, though they like the high treetops better. The only times I've seen Say's phoebes they've been on fence posts; I've found it pays to take a second look at any robin on a fence post—he's so apt to be no robin at all, but a Say's phoebe, sitting up even straighter than a robin, and duller colored. He has no robin accents of white on throat, eyelids, and tail corners; and the rusty breast has slipped, to be lower breast and underparts. If you watch for a little while, you will see the quick leap after an insect, and immediately the settling back to the same post; then you'll know for sure.

Magpies and ravens are anywhere, so of course you can sometimes find them along these fences. Magpies are a bit too much like mankind—unnerving, how they make you think of urban, sophisticated, smart-alec people. They walk about with a deliberate strut, cock their heads with an affectation of curious interest in what goes on, and comment loudly upon what they have seen. Related to crows and jays, they come between these in size and voice. Their clothing is quite stunning, a formal suit of black and white, always beautifully groomed, no matter what offal their feet and beaks might be in. The head is all black, with a saucy black beak and gleaming jet eye. Neck and breast are black, breast cut off sharply on a clean line from the shining white underparts. The back is black with a dazzling white U-shape on scapulars and rump, like a long white scarf looped down from the shoulders, a take-off priestly robe. The wings are black, patterned with white, visible only in flight. The feature you remember best is the very long tail, much longer than all the rest of the bird, black with a glimmering green sheen. If there's something on the meadows worth a magpie's attention, like the guts of a deer, a fresh seeding of oats or some other expensive seed, or a conspicuous hatch of insects, you'll find fifteen or twenty of these strutting rascals together, eating and talking and looking everybody over. It's like a cocktail party.

Ravens are in my study area only when something special brings them. Oddly, one time it was a heavy rain. Weather like that doesn't often happen in our valley. Showers, yes, when people and ravens can stay under shelter till the sun comes out again. But this was a two-day soaker, the kind that comes about once in three years. The ravens that daily overfly my study-mile apparently got so overweighted with water between the dump and the home place on Jakey's Fork that they had to stop to get rid of some. I saw several grumpy pairs that day, in each case two on adjoining fence posts: heads low, faces morose, bends of wings like high, hunched shoulders, feathers stringy and dripping. As the rain stopped and pale sunlight streaked through the clouds, they shook their wings, half spread them, and testily began grooming. Each feather was a struggle, had to have the water shaken off it all by itself.

Near the bottomland groves, house wrens find the bushes along the fences are the shelves of a well-stocked supermarket. The busy little mites almost mechanically flick eatables off the shelves (insects I can't even see) and into the shopping baskets down their throats. But they stop briefly every minute or so to ripple off a measure of song.

In August the young sage thrashers, just a-wing, come in from the sage terraces to the fences near the bridge. Every year there are three or four of the gray-striped youngsters for two or three weeks, then they disappear until next year. Day after day they chase each other around the willows and birches. There are so many insects around that eating seems less important than fun, while the babyish yellow folds at the corners of their mouths harden and darken into beak-stuff.

Western meadowlarks, who nest in the grass, like to sing from fence posts, but they are careful to pick bare ones, well away from any bushes.

One day in May, I saw a bit of high sky caught in a red birch bush in a fence row. Bluebirds are never in thickets, and besides this was deeper colored and too small. This was a fairly slow sort of bird, I could get a good look through the 8-power glasses. Such a beauty! The blue above gave way in a clean line to an arc of salmon color on breast and sides, and that gave way in turn to clean white underparts. There was a movement beside him. It was a plain little brownish tan bird with no stripes or accents. If Peterson's bird book didn't tell me so, I would never have guessed that this was the lazuli bunting's wife, or that the two were together for any reason deeper than coincidence. The book says they should be common here, but I've seen very few, so to me they are as exciting as if they were rare.

Every year a few green-tailed towhees shelter in the desert fence rows. Once in awhile in spring and fall one will appear on the ground at my feeder. They should nest in my study area, according to the books, but I haven't found any around in summer. They do nest every year at Trail Lake, eight miles away and 500 feet higher, on a steep canyon slope where a sage-and-grass ground cover is spotted with occasional small junipers. Perhaps the sage-and-grass around here isn't steep enough? Perhaps where the junipers grow here the other growth is too dense? I may hear one before I see him, a musical sparrow-song from a lower fence wire or right from the a ground, a sweet introduction and a burry ending; or maybe it will be just a loud, clear "Chink!" Green-tailed towhee is a poor kind of descriptive name. My private nickname is "red-headed green-bird". The olive-green is all over the back and wings as well as tail. The cap is bright red-brown. Underparts are gray, brilliantly ornamented by a spraying necktie of white with dark stripes, tied close under his beak.

In March, April, and May, again in September and October, the more thickety parts of the fence rows may be all aflutter with juncos, the quick little birds you seldom get a close look at. The small-sparrow size and white outer tail feathers may be all you get for clue, along with "Tut-tut-tut-!" remarks from behind screening leaves. A few other birds have white outer tail feathers. But meadowlarks, mourning doves, and solitaires are too big; and vesper sparrows are in the grasslands, not the thickets.

Vesper, chipping, and Brewer's sparrows do like to sing from the fences. Quite a few vespers and Brewers nest every year in my study area, but chipping sparrow likes it best just a little higher.

Last on my fence row list is the song sparrow. If the fence goes through his thicket, he'll be there, close to water, over and over in summer singing his lilting song.

Thickets

IN THE THICKETS ALONG THE RIVER I feel most at home. The species to which I belong must be ancestrally a child of the underbrush and open forest. Man's first home as described in the Bible was a garden. Anthropological research in Africa suggests much the same thing, but many million years before the time suggested by most interpretations of the records of Scripture.

And from those days to this there seems to be a categorical imperative laid upon man to make every effort to live in just such a garden. Look at high-level air photos of the Dakota plains, or the Sahara, or Yemen, or the Arizona desert: where you see the dark area of a thicket, there will be a house or houses. Put a group of pioneers on the high treeless plains: in a generation they will have created around each homestead a miniature forest and thicket. They will call the area "windbreak," "garden," "orchard," or "front yard;" but whatever they call it, it will be typically some kind of open forest with accompanying shrubbery or thicket. To people fleeing the city for the suburbs, it becomes a matter of status to turn the new suburban development into just such an open forest with shrubbery. Witness the arrival of Arbor Day as a national holiday. Witness the feeling that a ghetto is intolerable unless there are "parks" (that is, open forests with shrubbery).

Man tries hard enough to get away from this imperative. In any ecological survey you will read that man is the only species who inhabits all life zones. Man claims great honor for his species for his versatility and adaptability. You will find him, by virtue of his creative technology, deep under the ocean, climbing the highest mountain, setting up housekeeping on icebergs, inhabiting by the millions the barren cliffs of cities, walking on the sterile moon. But if he stays long anywhere, you will see he is not far

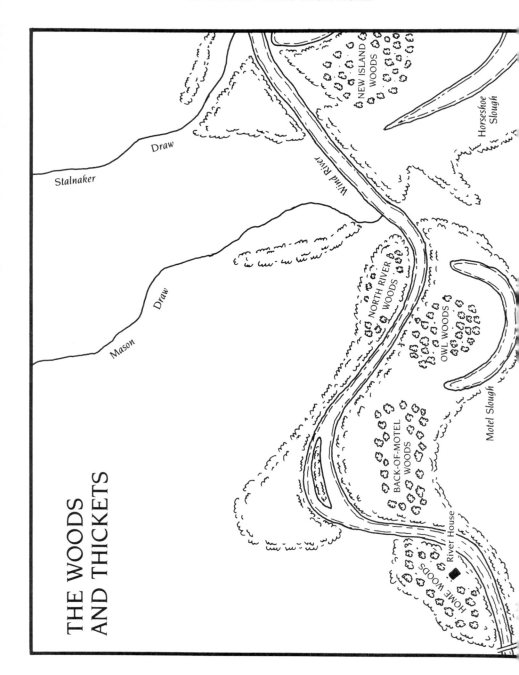

THICKETS

from some kind of thicket. His holiday resorts are almost always places of trees and shrubbery. He can't help it. Deep in his blood he needs them.

In my mile-across study area, there are twenty-eight residences and a fourteen-unit motel. Just beyond it to the west there are ten more residences. Most of these were built close to thickets; the rest are planting their thickets.

For purposes of this study I have divided the tree-and-shrubs into three life zones: thickets, forests, and yards-and-gardens. This is artificial. Life is one whole. But it's characteristic of my species that man learns better if he can somehow pigeonhole his knowledge. He is constantly surprised to find that his pigeonholes overflow, or lose their outlines, or don't contain accurately the knowledge that he tries to stuff into them. In this study, the constant complication is that each of the creatures (plant or animal) that I encounter is a free agent, an individual. He hasn't read the nature books to find out where he is supposed to be and how he is supposed to act. If he had, or if he were an inert substance instead of a living child of God, I'd have nothing to write about. It would all have been said already. Free choice is a characteristic of all life. If it were not so, there would be no surprises. Yet free choice is inhibited in so many ways—by our physical makeup, by our personality bent, by our social pressures, by our state of health, by the weather, by the food available, and most binding of all, by our own previous choices. This paradox—free choice but with countless binding limitations—is true not only of every human, but of each and every individual of all the forms of life that I encounter. So not all the creatures who like wild thickets will enjoy cultivated ones. Not all the ones that like short trees will also like tall ones. This variation is fascinating to me. Some creatures will be found in all zones, but each will like some spot more than another. Though I know pigeonholes are arbitrary, though I know free choice is going to make sport of my best efforts at organizing my study, still because I belong to mankind I have to do it!

The thickets are near the water: long thin stringers of brush along the river, hanging over it, leaning into it; or masses near and around the springs and sloughs; or dense growth at the outer edges of bottomland woodlands.

The bushes that make up the thickets are these: rocky mountain juniper, willows, young narrow-leaf cottonwoods, red birch, gooseberry, squaw currant, rose, silverberry, and sagebrush. Some other typical plants found in thickets are horsetails, willow-weeds, solomonplume, comandra, licorice, shooting star, penstemon, elephant head, and paintbrush.

The Rocky Mountain juniper is our mountain variety of red cedar. It is a tough, crooked little tree with stringy, shreddy bark, brownish-green scaly leaves, and lumpy bluish "berries" that turn out to be fleshy little cones. There is a scattering of them near the river clear across my study area: two big ones lying nearly prostrate on the riverbank just above Leseberg Island, two near our house, four or five in the Slough Woods, a couple on New Island. Old timers tell me there used to be many more, that these bottomland woods were once mainly cedars instead of cottonwoods. The cedars were cut for fence posts long ago. I suppose the half-mile of cedar posts up CM Draw are some of the corpses cut from here, gaunt and crooked skeletons that still hold up barbed wire after fifty years of use.

Willows are the most abundant trees making up the thickets, and they are surely the hardest to learn to recognize as species. You know the bush is a willow and no other, but which willow is the sticker. You recognize the willows by their whippy tough stems, the furry gray catkins on those stems in spring, the longish narrow leaves that are never opposite each other; and if you are in doubt you can take a chew of the bark—if willow, it will taste bitter, like . . . quinine? . . . or like . . . aspirin? Ha! the dictionary lets out the secret. Aspirin used to be made from willow bark before it was synthesized from carbolic acid. Well, well. Used (it says) to fight fever, headaches, and rheumatism. So that's what Grandma was after when she brewed such bitter tea out of willow bark and forced it down my brother and me when we had a fever. Aspirin, it was, all the time. Makes me wonder if moose and beaver ever have fever, headaches, or rheumatism—they live mostly on the willows. Nobody human uses willows for medicine any more—we are so weaned away from nature to pills. But willows are not endangered species. God willing, they will be around waiting when we need them.

When a new island raises its gravel out of the flood waters in late June, there is no plant life visible. Usually shoots of willows are the first to show up. Willows soon knit a fabric of lifestuff just under the surface, which keeps the new land from disintegrating. Most of the wood of our willows is underground. In my study area are no willows that could be called trees. The tallest reach no higher than fifteen feet; the largest stems are no thicker than my wrist, though the biggest individuals may have dozens of such stems. But, underground, the big roots are strange, crooked masses of solid wood, knobby and distorted, sometimes maybe six or eight inches across. A strange, secretive creature, this willow, raising thin feathery withes to bend in the winds, while solid massive growth reaches crookedly about underground.

Every book I have seen says *Salix* species are hard to tell apart, recommends that the casual observer not even try. I'm casual enough that I'll not try very hard. The sandbar willow is probably the commonest here. Its straight, smooth, dark red shoots are everywhere near water, on new bars, old stream banks, ditches, and around springs and sloughs.

Cottonwoods are much easier, since we have only one species in my study area. Some of my occasional walking companions think they're no easier, for our one species is the narrow-leaf cottonwood, and these friends say they can't tell cottonwood leaves from willows and can't tell the trees apart until the cottonwoods are big enough to be *real* trees. But the two have such different personalities. Willows are pliant yet tough, non-resistant yet whipping. Cottonwoods are much stiffer, have an awkward, angular air, and no real strength. Young willow stems are usually bright colored, red or yellow, while young cottonwoods are ashy or chalky or milky green. Every thicket in my study area has both. The young ones grow challengingly close together. As they grow older, the willow growth stays dense, but the cottonwoods space out. Many of them die, and only a few of the luckiest or most persistant become big fellows, well apart, shading an open forest floor. The leaves alone are certainly tricky enough as a device for recognition, being long and narrow, not too different from the leaves of sandbar willow, but they grow in little bunches, reaching out like fingers in all directions, where willow leaves march soldier-like in alternate steps up the stem.

Cottonwoods are well and truly named. In May, as the leaves are just starting, here comes the cotton. The trees flower and fruit in soft reddish caterpillar-like catkins. The cotton comes from the female catkins, which burst at every seed-pod down its length. Every seed has a cotton filament for a parachute to take it sailing far off down the wind. Sometimes the cotton piles up in drifts a foot deep.

The loveliest tree of the thickets is the red birch. In every season I am struck with joy at its unassuming beauty. Where it stands alone it grows symmetrically into a sphere, with a dozen or so stems three or four inches thick spraying out from one root. They are not straight but curve delicately, first out, then up, splitting into finer and more delicate shafts, ending in openwork lacy dark twigs, lovely against the rippling water or the gray winter sky. In May the swelling buds cover the dark lace with a transparent green veil, gradually changing as the leaves emerge into a nubbly openwork fabric, dark and light in a moving pattern. Each leaf is beautiful, the shape of a thumbnail an inch across, shiny, thin, veined in a fanlike pattern. In fall the color is a breath-stopping shimmery gold, becoming deeper as the leaves gradually blow off and the underlying dark lace comes into view again. The fruit is a little cone, the size of a little finger nail, hanging from the tip of a curving dark twig. The cones are usually abundant, accent points all winter in the lacy cover of the outside of the tree. Chickadees love the seeds in these cones. They're always fussing about them, filling the air with sweet little pipings. Seldom enough does a birch get to develop its spherical shape. Birches enjoy each other's company, like to grow close together in dense clusters. The interpenetrating fans of tough openwork make wooden walls, much more difficult to penetrate than the whipping withes of willows.

Gooseberries and wild roses are the brambles. They make the thickets tough to get through, yet they are such worthy local residents that we should be worse off without them. On the bad side, both gooseberries and roses have strong stems well covered with stout sharp thorns, often a quarter of an inch long. Gooseberry stems often look deceptively wooly, but don't be deceived. You can get painful wounds from them, and they are very tough on nylons and dress fabrics, and especially on sweaters. If you are gathering gooseberries, be sure to wear leather gloves. The berries themselves aren't prickly, but you do need protection to get near them. Rose thorns are stronger and more decorative, nicely spaced, light tan against dark red or gray stems.

On the good side, gooseberries and roses are pretty plants. Gooseberry blossoms are among the very earliest spring flowers, and the leaves are among the earliest bits of green. It's mighty heartening, as you grow weary of the long cold winter, to see the green in March on the slender arches of the gooseberry branches, and in early May to see those green arches decorated by small white flowers. Roses leaf out later and bloom later. In late June and throughout much of July the rose thickets are abundantly dressed up with pink blossoms as big as silver dollars and smelling like . . . well, like roses (is there anything else like that fragrance?). Both roses and gooseberries produce abundant and edible fruit, the only abundant edible fruit found in my study area. In addition, both come up readily, thickly, with sturdy independence, where the ground is scarred, filling in the tracks of bulldozers. Both make wonderful shelter for many kinds of small game, especially rabbits.

Gooseberries have an unexpected habit of growing a new bush where an old one touches the ground with the end of a shoot. New growth at this point goes in two directions: white, slender rootlets into the ground, a straight pale stem adorned with white thorns and bright green leaves into the air.

Weeks before other plants admit spring is here, sometimes in sheltered spots as early as February, gooseberry leafbuds will show green. The leaves themselves begin to unroll in late March and early April. Green gooseberry leaves heaped with snow are a startling sight but not at all uncommon. The small white flowers follow in a few weeks, and the berries are forming before wild rose even begins to bloom. The fruit of our species is deep purple. It's tart to the tongue, but cooks up with sugar into fine sauces and jellies. Bears love the berries. So do many birds, including waxwings, song sparrows, robins, starlings, and blackbirds. The berries begin ripening in mid-July, are at their prime in early August, and continue to appear in smaller and smaller quantities until frost. The bramble-crawling bird watcher is continually amused and impressed during this season by the quantities of bright purple droppings splashed about.

John Craighead says there are ten species of wild rose listed for the Rockies, but they are very hard to tell apart in the first place and they hybridize freely. So—three cheers—I shall be even more casual than with the willows, and speak only of one wild rose. In winter the rose thickets are dark red, spotted with scarlet—red for the stems, scarlet for the persistent fruits. They begin to be brushed with green in late May, though intermingling gooseberries have been looping their green arches for weeks. You can see the flowerbuds by mid-June, pink masses of bloom in July, and the earliest rose hips the first week of August. (Why that name, hips? Any connection escapes me.) They develop fast, and most years there is an abundant crop waiting to be harvested by early September, showing up bright scarlet among the green leaves. There are a good many customers: people making jelly of them, a clear pale jelly like apple jelly. Apples belong to the rose family; apple blossoms look a lot like wild roses; and the rose hips do look like tiny apples; they taste like apples, too. They have many other devotees. Robins, chipmunks, and squirrels are steady customers. A waist-high patch of roses dotted with fruit is good winter bait for bohemian waxwings. The birds will sit as near as they can to the tip ends of the branches where the fruit is, then stoop slowly, pull them off one at a time, and swallow them whole. Rose hips are fine for Christmas greens and winter bouquets. They are hardy, stick tight to their stems, don't dry up or shrivel. They look stunning with cedar, pine, or fir, and will stay brilliant for months.

Squaw currant is a close relative of gooseberry but much less assertive. No thorns. Not very common. It looks so much like a gooseberry shrub that you tend to avoid it on principle; you expect it to be thorny. It blooms much later than gooseberry—late May to late June—and has very pretty little flowers in pink sprays, long tubes with petals flaring around their mouths. It has bright red berries, but I've seldom seen them. I suppose the robins and blackbirds get them the moment they're edible.

THICKETS

One of the most abundant bushes of the study area is silverberry or mountain olive. It's a tough, crooked, shapeless little tree, with a rugged, dark stem and truly silver-colored furry leaves. I've learned that on June 18 I must start my walk with nostrils spread wide, to catch the first whiff of the year's loveliest smell—silverberry blossoms. When I catch the heavy, heady perfume (like gardenia) I look for the source and now can find it quickly. I was a long time finding it the first time, for I knew it had to be a large showy flower. I could hardly believe I was seeing straight when I discovered that all that sweetness came from a tiny yellow blossom no larger than the nail on my little finger, hiding between a gray silverberry leaf stem and the dark bark of the branch. Further looking established another surprising fact: the little flower has no petals at all. The little yellow points are stiff, woody bracts of a calyx protecting the stamens and pistil. I haven't yet found out who it is that fertilizes silverberry. There must be a story here. That powerful fragrance must be bait for someone, someone more useful to the plant than I. This is another of the mysteries that keep me walking my study-mile. Every once in a while I am rewarded by a piece of an answer; maybe tomorrow I'll be granted another piece. Whoever it is who is baited by the fragrance and accomplishes the fertilizing is mighty successful. The fruit is abundant. It deserves its name silverberry. It's a gray little olive with a rich sheen. In late summer, it seems as if almost every silver leaf has a silver berry hanging under it. Like gooseberries, currants, and roses, silverberries are popular with fruit-eating birds like robins, blackbirds, and waxwings. Since they're abundant and persistent, they are welcome food throughout the winter.

Shoots grow up vigorously from the spreading roots of adult plants. So one silverberry, starting from a pit like a small olive pit, planted by a robin or a chipmunk, will in a few years become a thicket. A few years later it will be a small forest fifteen feet tall. Though there are no thorns, the sweet-smelling jungle is hard to penetrate, for the tough crooked branches interlace in all directions.

Sagebrush is not often a part of riverside thickets. Where it is, it is sagebrush glorified. Craighead says there are 100 species, and that twenty are common in the Rockies. But it's just one species, big sagebrush, that is the emblem of the West. It grows all over the plains and dry foothills, common on southwest slopes up as high as 9000 feet; its fragrance is the very smell of the West. Typically, it's a ragged bush about two feet high, with one to several tough, crooked stems covered with stringy gray bark. The leaves are silvery-greenish at close range, covered top and bottom with a mat of fine hairs. Each leaf is a narrow wedge about an inch long, with three teeth on the wide end of the wedge. At a distance the plant is light-colored and nearly neutral in hue, responding to the color of the sky. It often looks bluish with sky-shine, sometimes lavender. A long spell of wet weather will turn it grayish green. It blooms in late August and September, signalling the end of summer; large numbers of flower stems shoot straight up all over the bush, bearing tiny yellowish tan flowers massed up the spike. The stems and seed containers stay until the next year's crop, giving a distinctive light fringe to each bush. The strong fragrance of the sage is all over the plant, in flowers, stems, leaves, and wood. It smells like a compound of turpentine, menthol, camphor, and ephedrine. Stroking the leaves, or brushing past them, releases quantities of pungent odor. It seems to be a volatile oil: whether green or dry, sagebrush burns readily and very fast.

This is a desert plant, spread all over the plains and benches where there's less than ten inches of water a year. It may fill the whole landscape, yet each plant grows alone, typically with its outermost twigs separated by a foot or so from adjoining sagebrush, fighting for its right to the small moisture in the ground. Too much moisture will kill it. Quite often, in opening up new land for agriculture, too much water is precisely the method used: open a new irrigation ditch, and just let the water soak into the area planned for a field. The next year most of the sagebrush can simply be raked up with a hayrake. It has drowned.

This sensitivity to too much water makes its occasional presence in the thickets seem strange. I can only attribute it to freedom of choice, or ignorance of the rules. In those spots where an individual sagebrush has decided to not let extra water kill it, it thrives and grows gigantic, up to ten feet tall, and with ridged, twisted stems of hard wood up to ten inches thick. It may shelter blackbird or robin nests. There are several of these giant sage-

THICKETS

brushes upstream and downstream from the bridge on both banks of Wind River. When Joe was first planning the bridge almost forty years ago, I was impressed by huge sagebrushes right where the bridge approaches would be. I managed to reach them with a hand saw before Joe got to them with an axe. I cut them down with care, and made several table and floor lamps of them. The most important part of preparing them, I found, was cleaning off the bark, stripping off the long tough strands of gray fiber. That bared the hard and shining wood with its curling stripes of white, tan, and greenish brown—beautiful, decorative stuff.

Since thickets are favorite haunts, I've put in considerable time and study on techniques of traversing them. There is a built-in paradox in human yearnings toward thickets, and the awkward build of the human foot. Man is really the only animal with hooks on the ends of his hind legs. Opossums, bears, raccoons, apes, and monkeys also walk flat on the whole hindfoot, it is true, but without well-developed heels the feet are much more limber and fold up more naturally when being pulled through thickets. Man alone seems destined by anatomy to be a stumbler. Part of the technique of getting through thickets is the trick of remembering those hindfoot hooks that are always with you. You walk rather slowly, stretching your toes and your foot at the ankle to slide past down timbers; otherwide you will trip over these hazards and pull them along with you. This footwork has to become instinctive, because your conscious efforts have to be to protect your face and especially your eyes. You learn to keep your head low, left arm well ahead of your face, right hand spread in front of your eyes. You can usually plow through a thick willow growth like a ship in heavy seas if you use considerable brute strength. Red birches require more body English, more sophisticated gestures of body and arms. You push through young cottonwoods with straight forward determination. Silverberry jungles may force you to your hands and knees. Finally, you mostly have to give up and go around dense rose and gooseberry brambles.

But the thicket crawler isn't always faced with pathless brush. Paths when they exist are surely the best way to get through. And paths there are. The easiest paths are the fisherman's, always available as close to the very edge of the water as possible. Wherever bites of grass tempt them, horses make ways through, ways which are plenty wide enough for you, and which force you to stoop a little. Deer trails are narrower and not so tall. There are paths trampled by raccoons and bobcats, but following them is likely to be a hands-and-knees job. Even the densest thickets are well penetrated by trails of cottontails, pack rats, squirrels and mice. But even the most persistent and smallest human bramble-crawler is too big for these.

The path builders are the most conspicuous of the mammals, but there are others. The little brown bat on summer evenings may whizz past your face. His incredibly sensitive sonar system lets him make intricate maneuvers in the midst of the web of twigs. His headlong flight looks as sure as it looks heedless. They must break their slender wing bones sometimes.

One September 12 I myself broke the humerus of a little brown bat. I was bothered by the swish and the tiny, high-pitched cries near my head in the dark. I hit out hard and knew I hit him. I searched him out and found him huddled in a cranny, one wing neatly folded, the other a crumpled rag. The upper bone of his left arm was broken just above the elbow. I was smitten with remorse when the little fellow stood up, nervy as they come, and chattered his teeth and squeaked at me, for I knew I could do nothing for his injured wing. I had to kill him. I did it as quickly as I could, pinching his lungs firmly between my thumb and forefinger. Afterward I made a drawing, marveling all the while at the tiny precision of his structure, so much like ours that the differences are only more brightly illuminated.

He looked like a mouse with wings, a tiny little fellow with glossy brown fur. The skin of his wings was like the skin of the back of my hand, a darker flesh-color, not furry, with an intricate pattern of blood vessels between the inner and outer layers. The muscles and tendons of the arms and legs are so slender that the bones are clearly seen, almost bare under the skin. The arms have a short humerus, a very long radius-ulna, three enormously long fingers supporting the web of skin, and a small thumb sticking out free. The wrist structure seemed complete: I felt sure that careful dissection under a microscope would show eight wrist-

THICKETS

bones just as we have. In the hind legs, the femur and tibia-fibula are turned sidewise, in a position we could never hold. They are built firmly into the web, the femur short, the knee near the edge of the body fur, and the five-toed foot just outside the web. Ten tiny tail bones support a leaf-shaped membrane between the hind legs. The tail bones curl inward, shaping this web into a cup. (I read that the little brown bat female bears her one baby as she hangs upright from a roost, and that the baby falls into this cup and crawls up the mother's fur to her breast to suckle. What a sight to watch! Something else to look forward to.) Each ear is a big delicate scoop, nearly the size of his skull. Sticking up from the base of the scoop is a tiny wand. We have like projections at the bases of *our* ear-scoops, but ours are shaped like the halves of pennies, not like wands. The engine of this tiny living machine, the power that makes it fly, is the pair of pectoral muscles over the front of his chest. This muscle mass is enormous compared to the rest of the bat. It gives him an exaggerated chesty look, overstuffed, and emphasized by a pair of dark furry epaulets at the bend of each shoulder.

One time an otter put on a great show of people-watching. My friend Betty and I first thought he was a muskrat, swimming in Horseshoe Slough, then realized the "muskrat" was just the head of an otter. He swam straight toward us, watching us. As he passed, he exaggerated his undulations to make a muscle-shining show. He crawled out on a rock and turned over languidly, belly-up to the sun, making sure we were looking; swam back to a few feet from us and stood up straight in the water, till we could see his navel—feet had to be paddling furiously, but no sign of it above water. Then he dived, caught a trout and ate it, before slipping into thicket-shadows.

Several other members of the weasel family have meant something to me. Once by the New Slough I met a long-tailed weasel eye-to-eye, only a few feet apart. He was near the top of a sagebrush on the slope of the highway fill; I was lower down the slope; our eyes were on a level. A heavy April snow was piled high all over the bush, with small V-shaped dark spots like windows here and there. I was attracted by a light shivering of the whole bush, making me suspect an inhabitant. I stood quietly looking at it. Weasel looked out at me, filling an entire V-shaped window with his head and neck. I suppose he was climbing about in the twilight on the sheltered inside branches, maybe chasing and catching mice and making the sagebrush shiver. He heard me coming, then was puzzled by the silence, and had to look out a window to satisfy his curiosity. He was so startled he seemed paralyzed for the space of several seconds. He was in summer pelage already, and very brilliant against the unseemly spring snow—bright red-brown, flat-topped head, small round ears, creamy white throat and chest, black eyes and tip of nose. A quick spark flashed in his eyes, and the window was dark and empty.

Skunks are thriving. Several times in the twenty-plus years of these observations Buttons and I have seen these big black-and-white striped weasels at close quarters. Most of the sightings were painless and harmless to us both. Vivid memories cling to the one that wasn't. It was early September. Buttons and I were following a fisherman's trail next to the willow brush, Buttons leading, as usual. I heard the happy burst of yelping that signals a rabbit or something and saw the delighted leap under the willows—the

THICKETS

leap that turned in the air into a somersault, a backward scramble, a frantic clawing at her neck, rubbing of her head on the ground. In a second or so the cloud of dreadful perfume gave me the whole story. I didn't need the accompanying view of the black-and-white creature in the deep shadow, watching beady-eyed, patting his feet. Apparently the skunk was sure of himself by then, confident of victory, or he'd have been in his more threatening tail-first attitude. Or perhaps he had shot his wad and was out of ammunition. It was quite a wad. I was well out of line for the direct hit, which caught Buttons on the front of her neck just below her face, but the indirect load I got was plenty. The few seconds it took us both to get out of range were long ones. We were reeling and gasping as we pushed out of the thicket into the clean west wind.

Buttons smelled awful. I knew she'd not be allowed in the house for a week; what I didn't know was that I was in the same situation. Since I wasn't hit directly, my lungs were well cleaned out by the time I got home, and I was all set to tell Joe the story. Which he didn't need. I wasn't even inside the door when with emphatic gestures he put me back out and closed the door against me. "Out!" he called, laughing and gagging at the same time. "Undress on the porch and throw your clothes out in the yard. Then I'll let you in the house if you'll run fast to the bathroom!"

A half-hour soapy shower and shampoo got me to where Joe said I smelled respectable. Three hours of September west wind through the house made it habitable again. Overnight on the clothesline lightened the flavor of the clothes to the point where I could stand to carry them to the washing machine, and a good laundering did the rest. Buttons had her week outdoors.

I hope I never suffer a direct hit!

I guess I am really chicken. I dream of all kinds of investigations into the lives of the brother and sister species who share my corner of the earth. As I admire wildlife movies, about cougars and brown bears and wolverines, I see myself in the role of the cameraman. As I read reports of field scientists, I imagine I am doing the same things. But when the moose or bobcat or coyote actually appears before my eyes I am suddenly circumspect. I become willing to do my observing from some distance, such as from inside a jeep or even from behind the windows of my house. It's actually good luck rather than courage that has put me in position for closeups once in a while.

Quite common in the thickets is the frisky little chipmunk. In the eastern states you could get right into telling about his endearing ways, but in the West you have to get bogged down first telling who he is. There's only one species all through the East, but there are sixteen species in the West, and many more subspecies or races. I've had to do quite a bit of studying. It turns out that the species in my study area must be the least chipmunk (*Eutamias minimus*). He is smaller than the eastern chipmunk, seven to nine inches instead of eight to twelve, and much paler and grayer in background color; his black and white stripes show up more, especially on his cute little pointed face.

I've had no luck in taming the chipmunks. They remain very shy, though they live merrily in my woodpile and on my bird feeder. Very curious, they announce themselves with a birdlike chirp that brings Buttons running, upon which they disappear abruptly. Rarely, I get to see a little of their domestic life. Once, two of them were having a banquet on blue cedar berries. They were full and happy. One quick demonstration of affection sticks in my mind. She was spread out, head first, on the lower trunk of a juniper. He was coming headfirst down it, with a blue cedar berry in his mouth. When he got close, all at the same time he threw away the berry, seized her face in both hands and gave her a quick smooch on both cheeks. Becoming aware that he was being watched, he jumped away from her, ran up the trunk to a small branch from which he scolded me; then he lost himself in the foliage while she disappeared among the roots.

They are very good climbers, though they spend most of their lives on the ground. I see them way up in the slender black lace of the outer twigs of red birch, eating the small cones. They like silverberries, too, and climb fearlessly for them. I've seen them on branches bending under their weight right over fast open water. Once I surprised a chipmunk sleeping in the sun in the top of a squaw current.

Mice are quite common but by no means over-abundant. When there is a little snow cover, I see the delicate embroidery of their

THICKETS

tracks here and there near brush. According to the guide books, these should be harvest mice, kangaroo mice, grasshopper mice, pocket mice, and deer mice, but deer mice are the only ones I know. I have caught them in mousetraps. I have killed them with poison. I have met them in clothes closets and food cupboards. I have found their nests and young families. I have learned to admire and respect them. They are alert, intelligent, and stubborn. They have a great capacity for joy. And they are beautiful. They are soft warm brown above and pure white below. They are pop-eyed; the big dark eyes stand out from their heads. The big scoops of their ears are brown-furred outside, dark pink skin inside. Like their bodies, their tails are brown above and white below. Their fine tiny hands and feet are pinkish white.

Years ago, an encounter with deer mice taught me a powerful lesson in community relationships. Joe and I had just set up camp in the mountains with Button's predecessor in our family, a black shepherd dog named Monkey, a dog who had given abundant evidence of a happy interest in the welfare of all forms of life.

Well, the first morning in this camp in the mountains I was wakened in early dawn by a scrambling of little feet near my head, and a chorus of small squeals. I turned my head very slowly and saw two deer mice dancing on the tarpaulin beside my sleeping bag. They were scurrying round and round in a circular pattern the size of a plate to the accompaniment of tiny shrieks. At intervals they would run in to the center from opposite sides, stand up facing each other with fingers touching, drop to the ground, back to the circle, and scurry after each other again. It looked like a happy dance. I very slowly turned my head to see what Monkey was doing. He was standing up in his place at the foot of Joe's sleeping bag, craning his neck to see over me, and smiling broadly. My motion or the growing daylight scared the mice, and they disappeared under the tent wall.

When we checked our supplies, we knew why the mice were so happy. They had found our oatmeal, bread, and leftover pancakes and had a big feed. We were disgusted at the mice and mad at them. We should have been mad at ourselves for leaving temptation in their way. Instead, we set a trap.

Deep in the night we were awakened by the snap of the trap

and the loud squeak of a mouse in pain. The flashlight showed one of the mice caught by the hind legs, and its mate trying to pull it loose by tugging on the skin of its neck. Giving up on that, the free mouse ran around, braced itself against the trap spring, and tried to push its mate out. As Joe and I spoke, the free mouse jumped out of sight into the dark. I looked at Monkey, curled up at the foot of Joe's bed. He was watching the mice, his face wrinkled and sorrowful.

Joe knew he was going to have to get up and kill the trapped mouse. He tried to get Monkey to do it for him. He said firmly, "Sic 'em, Monkey!" Monkey turned on him a face filled with absolute horror. He stood up stiff-legged and stalked out of the tent. He didn't come back for many hours, not till long after breakfast, and he hadn't forgiven Joe then. Of course, long before then, Joe had put the little mouse out of his agony. Monkey told us plainly that time that Joe was Caesar and not God. He as much as said, "Joe is my Caesar and I do as he says, usually without question. But when it comes to hurting God's children, God is love and love comes first!"

This experience considerably changed our attitude toward dogs, mice, and men. At the very least, it caused us to depend more on mouse-proof containers than on traps or poison.

Of all creatures that belong in the thickets, the most at home is the cottontail. Surely from his point of view it was for him they were designed. The thickest part of the thorniest brambles is his refuge. He seldom ventures more than a few jumps from shelter. According to the field guide, the species in my study area is the mountain cottontail, a different species from the one I knew back east. I was surprised to learn that, for it sure looks like the same bunny to me. It averages a little smaller than the eastern cottontail, and the book says it is a little paler. I think that the ones that live in the sage are paler than the cottontails in the thickets, but I'm not going to kill a bunch of bunnies to find out for sure. Cottontails are active mostly at night and spend most of the daytime hours snuggled into a self-made hollow in long grass and among abundant thorns. Gooseberries and rose patches are their special delight. Since my walks are usually from dawn to soon after sunrise, Buttons and I often surprise bunnies as near to the open as they ever come. This is Button's prime joy, the real reason she looks forward to walks: she is always hoping to jump a rabbit. When she does, even now in her sedate old age, she gives a loud cry, then a long series of happy yelps as she scrambles off at a great rate, usually in the wrong direction. She has never caught a rabbit, and the chase is never long. Rabbits are clever at changing direction in midflight, and even if she got started right, Buttons gets puzzled at the first evasive action. I can't laugh, though, I can't imagine myself even starting a rabbit chase with milky cataracts over both eyes; when she gives up and comes back I must reward her with a straight face and a "Good try, Buttons."

Mule deer, whose real home is in the desert, belong in the thickets, too. The brush in my study area shelters perhaps ten deer off and on. I often see their tracks, and every once in a while Buttons and I encounter some of them. They come from the Badlands to the north and the moraine hills to the south for their daily drinks of water. They play a large part in keeping the thicket trails beaten down as they thread their way unseen through the underbrush.

It was hard to convince Buttons that she must not chase deer, but I think she has learned. After a wild yelp and a few yards bolted, she stops and waits for me to call her to heel. Not all dogs know that lesson. One of the troubles that game wardens have is coping with packs of Dubois dogs that run deer, sometimes killing them. So far, it hasn't happened in my study area. The only experience I've had in that direction concerned not mule deer but another member of that family. An occasional companion on my walks is a little lady who owns a big, gentle part Labrador. One early spring day in back of the motel, we heard a wild outcry that told us her Snuffy had started something. The sound was coming our way as we threaded a narrow deer trail. Suddenly, here was a young *moose* bearing down on us, closely pursued by Snuffy, who was yelling, "Hey, Ma! Look what I found!" or something like that. With no hesitation at all we got off the trail into heavy silverberry and rose growth, which I know we couldn't ordinarily have dented. The moose went on by, and Snuffy, having made his point, stopped with us.

Small birds in general take to the thickets. It follows that their predators do, too. Chief among these is the sharp-shinned hawk. Sharp-shin is the smallest of three fierce hawks called the Accipiters: goshawk, Cooper's, and sharp-shin. Goshawk is as big as a raven, Cooper's big as a small crow, and sharp-shin about the size of a robin. I see them all sometimes. They are all swift killers. Their wings are short and round, tails long and held closed; it is a build designed for short flights and for evading at high speed such obstacles as the millions of branches in thickets. All three do hunt in thickets, but the larger and heavier Cooper's and goshawks have more room to maneuver in open woods and open country close to woods. Sharp-shin really belongs in the thickets.

They are fiercely handsome little marauders. The adults have sleek blue-gray caps and cloaks, barred rusty breasts, white throats, white belly, white tail bars, and flaming red eyes. The immatures are barred brown and tan all over. In the years of my study I've had seventeen encounters with them; this means, I am sure, that they've encountered me at least three times as often, but I've not seen them. They are hard to see; they spend most of their time sitting still as a branch among the branches, waiting for a chance to strike. Let a little bird venture just a hairline too close, and there's one swift swoop and it's all over; the yellow talons have pierced and crushed it. One more meal for sharp-shin.

In summer the thickets are good nurseries for insects, so some flycatchers can be expected here. The flycatcher style identifies them: each one spends more time sitting still than flying, sitting very straight, body and tail in a nearly vertical line. When he spots a nearby insect in the air, the bird flies swiftly to it, sometimes almost straight up, takes the insect neatly in the air, and returns to the takeoff point for another wait. Five flycatchers are common to the thickets I watch, but only in summer; I suppose because that's the time of the most insects. Eastern and western kingbirds hunt the thickets as well as the open country. The tiny fellows, grayish Traill's and yellowish western flycatchers, are found in the willow and birch thickets close to Wind River.

Magpies may be seen in all kinds of places—deserts, barnyards, lawns, highways, fence rows, big woods—but thickets are their home. Like people, they are bold, quarrelsome, sociable, alert, curious, and talkative. Sometimes the thickets are loud with their comments and conversations. There are a dozen or so magpie nests in the thickets I am watching. The nests are ragged collections of sticks as big as bushel baskets, looking so carelessly put together they should fall apart in the course of one season, yet they last for years. They show up plainly against the sky in the winter, at various heights from about fifteen to about thirty feet, mostly in middle-sized cottonwoods. The birds may build new nests in the spring or refurbish old ones. They may roof a nest or leave it wide open, the big mass supporting a platform with a dished hollow lined with rootlets or grass or both. One I saw was carefully protected by a lot of prickly pear cactus tucked around the rim. Made me wonder if last year a pack rat had gotten magpie eggs from an unprotected nest, and that magpie was not going to let that happen again.

One March morning after a heavy, wet, all night snow I started about 6:30 to walk the Lower Pasture. The brush looked strange, all ludicrously bowed with a great weight of snow. A birch tree was flattened till it was no higher than my head and twice its normal width. I was just about to pass it when it exploded, quite literally. The branches all shed their snow and whirred up to normal height, and with a loud sound like tearing silk two dozen magpies spread their wings at once and leaped into the sky. The air was full of snow and magpies. They must have been delightfully snug and warm, nestled side by side on the network of weighted branches under the roof of snow. Twenty-four bodies at a temperature of 110° each, and with snowy insulation to keep the heat from leaking away. But daylight was beginning to show here and there in cracks in the roof and walls. Magpies were beginning to open black eyelids and sleepily move black heads. Squuunch! Squuunch! Squuunch! The sound of boots in the snow! Getting closer! So ... R—R—R—R—Rip! In one concerted action, the bedroom was empty, the roof torn to sheds, the birch tree remade into the shape of a sphere, and the flock in the air.

From large black-and-white to tiny black-white-and-gray—next come the chickadees. There are two species here, both of them little fellows not as big in the body as your thumb, but more alive than anybody. The black-cap is much like the one I knew back east, only his tail is a little longer. White breast, gray back, white cheeks, wide black tie, delicate warm tone on the flank feathers that cover the edge of his folded wings. The mountain chickadee lives at a higher altitude in the evergreen timber, but I see him occasionally in the river thickets in winter. When he is around, I recognize him by the sound before I see him, because his "Chickadee-dee-dee" is much lower and huskier than black-cap's sweet, clean-edged phrasing. So I know in advance that I'm to look for a tiny bird with black-and-white striped head: there is a wide white eyebrow stripe cutting along the black cap on both sides. Also there is no pinkish brown on his sides; the bird is all black, white, and gray.

I cannot quite believe in chickadees. Their cheery bounce in bitter weather is not only unreasonable, it's incredible. From the angle of mechanics it doesn't make sense. If you were looking from God's viewpoint, imagine the problems of building a machine with these specifications:

> Length: 4½ inches, including 2 inches tail.
> Must be insulated for 40° below zero.
> Must include all services: digestion, elimination, reproduction, blood circulation, lymphatic system, endocrine glands, nervous system, bones, muscles, skin, and aerodynamics.
> Must maintain constant temperature of 110°.
> Fuel supply: insects found on tree bark (plentiful and lively in summer, dormant and hidden in winter).

The miniaturization of all services is beyond belief. I *know* these things but I can't believe them: the little dee-dee is equipped with a brain, spinal cord, and complete nervous system much like mine; a complete thermostat-controlled circulatory system like mine, with a four-chambered heart and a network of tiny arteries and veins tied to lung-stuff like mine; a digestive system complete with tongue, esophagus, stomach, liver, pancreas, intestines, kidneys, bladder; a muscular system completely parallel to people's; a full set of endocrine glands; feathers of an astonishing complexity instead of body hair, detailed to the point that the miniature eyelids are equipped with eyelashes which magnified can be seen to be feathers, too.

In summer wrens replace chickadees as the liveliest birds of the thickets. Three species are on my list: Bewick's and the winter wren just barely, and abundant records of house wrens. You hear them before you see them. Over the years that I've watched these thickets, I've heard them in the spring as early as April 8. What I hear is a clear loud fluting melody, interspersed with a bit of hoarseness: "Tuk-tk-tk-tk-tchk-churrily-churrily-warble!" It's short, repeated, and much too big for the little throat that's putting it out. If I am too close, what I hear may be a buzzy, choppy scolding, directed straight at me. While the chickadees are friendly and curious, the wrens are domestic, possessive, and volubly defensive. They are even smaller than chickadees. If they weren't so muscular and talkative you would hardly ever see them, for the little fellows are colored to disappear into the tree bark where they live. But they are so lively and unafraid that all through summer you are much aware of them: tiny brown creatures finely barred with darker brown and black, narrow longish bills that seem to droop just a little, a jerky little tail that often stands straight up. Wrens must be the most efficient insecticides of our woods. They are always at it through the daylight hours, taking insects from the air and from all over the leaves, branches, and bark. Mostly they take small stuff, like midges, mosquitoes, and insect eggs, but I've seen them with all sizes of caterpillars and with moths up to three-inch wingspread. Their main foraging grounds are the riverside thickets, but their homes are in holes in the big cottonwoods or in people-made bird houses. They will show up later in my narrative.

Catbirds lurk in the thickets. Lurk is a good word and fits their secretive ways. On the deep-shaded ground or in tangled branches, they seem to be always trying to shrink into invisibility. Their efforts to disappear are abetted by their coloring, foggy gray darkening to black. There's a contradiction: under the tail is a patch of bright chestnut. This is such an odd spot for the bird's brightest color that it encourages speculation that it, too, may be camouflage, designed to make a marauding hawk think that's the head end. The only thing accomplished thereby would be to have the hawk feel so silly at the last moment that his aim would falter, deflecting just enough to allow a quick dodge behind a branch.

Catbird's voice belies his color and mannerisms. It's rich, loud, flexible, and melodious when he sings his own song. While singing he forgets about lurking. He may even be skylined at the top of a rosebush, with head feathers raised, head high, throat swelled, wings quivering. I have always heard them before seeing them. The call is a soft miaow, like a cat—hence the name. Besides the call and the bird's own song, each one has quite a repertoire of imitations. I have heard catbirds doing robins, yellow warblers, meadowlarks, and sage thrashers, sometimes confidently, sometimes experimentally as if practicing. They are good insect killers; it doesn't take many days for one pair to clean up completely a big tent-caterpillar colony. From what I have seen, I think they concentrate on crawling rather than flying stages of insects. They are here for a short breeding season. My records of arrivals are all late May or early June. I guess that on account of their cautious ways they may have been here several days before I knew it. They leave early, too; September 23 is the latest date I have.

Most of the species of thrushes are thicket-birds. The best known of all the thrushes, robin, forages confidently there, as well as in the open woods and in the yards and gardens. I call him a thicket-bird, though any robin observer can put up a good argument that robin belongs to any habitat. I'd have to admit that you can find robins in lodgepole pine or Douglas fir forests high above my study-mile. They spend considerable time in the hayfields or in the dry sagebrush desert. I've even seen them above timberline on the tundra. Once in awhile a robin in our area will spend the whole winter in our valley in some sheltered spot by a warm spring. (The main bunch must not go too far south, for the time they allow for wintering is short. They are among the first comers in early March and seem to be all on hand by April 15. Most have drifted away by early October, but some will linger way into November.) Remarkably hardy birds they are, and with richly varied tastes in homesites. Although they like to stay in thickets, their mud nests are too heavy for slender branches. They need log ends, board shelves, or crotches of big trees. But still, with all this variation in habits, they do live much of the time in the thickets, running over the ground among the willow bushes or singing from the birch trees. During the chancy time of adolescence, the smarter young robins use the thickets for study rooms and gymnasiums while they are learning to fly.

Does the human monster exist that will hunt robins? I find it hard to believe. (I've heard of blackbird pies but never of robin pies.) Their serenity, aplomb, courage, gentle ways, cheerful voice, and bright color must surely endear them to every watcher. And yet, in some places, DDT has decimated robin populations. Though no one would deliberately engage in robin-killing, spraying for insects with DDT does the same thing. Robins will eat dead or dying sprayed insects and soon die themselves. Twice these woods I watch have been sprayed to keep down the mosquito population for the benefit of guests at the Red Rock Ranch Motel, and each time the thrush and flycatcher populations have suffered. Fortunately there have been successive unsprayed years during which they have made a comeback. People may sometimes begrudge robins strawberries, but they should in turn give thanks for robin's gift to people. Robins are great insect killers, doing an enthusiastic job of making the world livable for people during insect seasons.

They have plenty of non-human enemies. One April day I found a little pile of feathers by the big willow near the bridge. Enough were burnt orange in color to make sure the lot came from a robin. There was one whole wing, which had been torn raggedly from the torso. The aerodynamic feathers (primaries, secondaries, thumb feathers, primary and secondary converts, and scapulars), only a little disarranged, were dark gray above with light gray edges, light silvery gray below with burnt orange inner coverts. A beautiful, functional form, now useless. I took it home to put in my journal, as I wondered if it was a mink, an owl, or a hawk who did the job.

The shy hermits are close relatives of the robins. Their build is similar and they are nearly as large. There's the same innocent look: that has to do, I think, with a forehead-like drop from the top of the head to the beak and eyes wide apart. There's the same habit of running fast for a dozen steps, then stopping and looking around, then running again. Diets of both species are the same, with hermits as anxious as robins to find worms, caterpillars, moths, and gooseberries. But hermits, unlike robins, are cautious in my woods, staying in the deep shade of the thickets in May and in September. Their coloring is much more of a camouflage than robin's is: soft medium brown above, enriching backwards to a bright cinnamon or fox-colored tail, light tan breast with a sprinkle of small spots, big round brown eyes. Hermit is mostly silent in my woods, but he is the beautiful singer of the fir forest at an altitude of around 8500 to 10,000 feet. On back-pack trips into his chosen habitat, around sunrise or sunset, I see him sitting in plain sight in the tip of a tall tree and hear him pouring out a liquid cascade of magical song.

I see Townsend's solitaire sometimes in my thickets, but only in winter. Like hermit, he lives high up in the evergreen forest, where his long warbling song is one of the rewards for exploring. He is a hardy bird, only occasionally descending as low as my study area. When he goes down the mountain sides, away from snowbound forests, he prefers to linger in cliffy canyons with Rocky Mountain juniper growing on the ledges. I can count on seeing a few each winter on the steep slopes of Torrey Canyon among the junipers.

A tiny sprite of the willows is the ruby-crowned kinglet. In late April and early May and again in September, the brush may be jumping with the little fellows. They are on the move every second and even smaller than warblers, so small they resemble willow leaves in the wind. They are busily looking for insects, caterpillars, and insect eggs, and are apparently reaping a steady harvest. The field marks are this tiny size and constant motion, a greenish gray color, two white wing bars and an eye-ring, and a narrow red crest. The red you can't count on, for only the male has it, and it doesn't stand up unless he is excited; otherwise it lies flat and the gray head feathers partly cover it up. He has a clear call-note: "A-wheat! A-wheat!" and a lovely rippling song. I've seldom heard the song in my study area, but it's a regular part of the June morning chorus in the Douglas fir and Engelmann spruce forest high on the mountains.

The elegant waxwings come here sometimes. I value them the more because I can't count on them. There are two species, cedar and Bohemian, and occasionally there are some of both in the same flock. Cedars are more likely in summer, Bohemians in winter. One summer a pair of cedar waxwings nested on a horizon-

THICKETS

tal branch of cottonwood just above the willow brush at our cookout, and we grew used to the soft whiffling lisp that was his song. Waxwings are almost the size of robins, Bohemians a little smaller, cedars a little smaller yet. Cedar is brown, Bohemian grayish brown. Both have proud high brown crests and yellow bands across the ends of their tails. I think Bohemians are rounder and cedars are sleeker, but that may be an illusion because I see Bohemians in cold weather when their feathers are fluffed out for insulation. You look under their tails for the positive field mark: cedar undertails are white, Bohemians are rusty red. Another good one: cedars have yellow bellies, Bohemians gray. The wax of the waxwing's name is a series of small scarlet droplets like sealing wax on the ends of the shafts of many of the secondary flight feathers, the feathers that grow out of the edge of the ulna bone. When the wing is folded, the tiny brilliant beads add a handsome decoration if you're close enough to see it. There's an odd factor of individual difference in that not all waxwings have wax, nor is every secondary feather so equipped on the ones that have some; and this variation seems not to be related to age or sex.

Shrikes are uncommon but not unexpected visitors, winter or summer. They are songbirds gone wrong. On the species list of my area, shrikes come between waxwings and starlings, but in their habits they should be listed with the hawks. Hawks are respectable predators, shrikes are disreputable predators. The reason we feel that way about them seems to be that hawks are true to their physical structure, shrikes are false to theirs. Shrikes as songbirds have a rather sweet warble. As predators they have no talons to hold, no strong hook to their beaks to strike or tear. Both our species are rather trim and pretty birds in gray, black, and white, reminding me a little of Clark's nutcrackers or gray jays. Both species are about robin size. Northern shrike, the winter one, is a little larger, a gray bird with a big white head cut by a jaunty black mask across the eyes. The wings are black with an irregular white patch on each. His gray breast is very finely barred with wavy darker gray (you have to be awfully close to see it). The summer shrike, the loggerhead, is a little smaller bird with a little bigger head. The markings are about the same, except there aren't those almost invisible barrings on the breast.

THICKETS

Shrikes aren't especially birds of the thickets. Usually you see one sitting on the top of something; it may be a tall tree or a short one, or even a greasewood or sagebrush. But since they are predators they come where their prey is, and many of the available small birds shelter in thickets. On a November day I heard a harsh, weak scream as a northern shrike made a gliding swoop from the tip of one cottonwood to the tip of a shorter one near the river just above a birch bush. A pair of black-caps who'd been cheerfully scouting the twigs of that birch took quick shelter deep inside and became very quiet. But after a few minutes they tuned up again; still very conscious of the marauder, looking at him all the time; they seemed to be teasing him from a dense shelter of twigs. His reaction was to shift about as if uneasy and frustrated. Three weeks later I saw a shrike (I wonder if it was the same one?) chase a downy woodpecker from near the bridge into the cottonwoods near our house. The downy ducked, turned, dived, and got away into dense thicket.

One October day I saw the smaller loggerhead shrike kill an Audubon warbler. The bright little bird left the nearby shelter of silverberries to get a drink from the Wind River. Now he was briskly grooming his pretty yellow and blue and white feathers, spreading one wing at a time and with his beak reorganizing each pinion. When he saw the shrike swooping low across the river, he tried to make it back to safety, but shrike was too fast. Coming up from behind he covered the warbler with his wings and bore him to the ground by the weight of his body. Standing on him, he killed him by beating with his beak on the little bird's head. It took several tries; I could tell by the continued fluttering of the warbler; it seemed as if the shrike was not very strong. But the warber finally died, the shrike standing on the little body until it was still. Then he straddled him and crouched, gathering up the body between his legs and into his belly feathers, and flew with him across Wind River. I suppose this maneuver was indicated on account of my presence; he needed privacy for his dining room. Whatever his reason, he was hidden in willow brush.

There are the starlings. I'm tempted to ignore them; almost everyone knows them only too well; maybe the old theory will work that if you ignore the problem it will go away. Starlings are not so obliging. They spread rapidly across the continent and found their way into my area about 1955. Starlings' tastes in habitat are much like people's, which goes far to explain why we see so many. Travel to areas away from people (like the sage benches, the evergreen forests, the high tundra) and you'll be away from starlings, too. They are usually around houses or barns and in the thickets and woods nearby. From here they sally forth in noisy crowds to meadows, sloughs, or riverbanks where the feeding is good. They are omnivorous eaters, going after insects, fruits, berries, or seeds with equal relish, delighted to find a meaty bone or a hunk of suet. As individuals they would not be objectionable neighbors, but they are aggressively social and have discovered what can be accomplished by mob tactics. They tend to take over completely a patch of woodland. Other birds, like robins, wrens, and flickers, give up and move away out of sheer dislike of the starlings' presence. Flickers, for instance, are much bigger and fiercer than starlings, but I saw a pair faced down by two of them. The flickers had drilled out a good hole above my clothesline in a big cottonwood trunk and had been carrying nesting material, but the pair of starlings saw an opportunity when the flickers were away. They threw out all the flicker nest lining and carried in their own choice of grass and feathers. While they were about it the flickers came back. They took a look-see, turned up their beaks, and went away from there. I haven't seen starlings actually scrapping with other birds, but wherever they are, there seems to be a sanitizing bird-free belt around their flock.

It's a pleasure to turn from starlings to the ten species of warblers I've found in my thickets. Most of these bright little birds live much farther north or higher up, and we see them only on a few hectic days in May and September. The warbler days are exciting ones. If there has been a cold snowstorm in late April or early May to hold back the migration, the first warm day will bring such a flood of them that you will nearly go crazy trying to see them all. They have to belong to this list: yellow-rumped (Audubon or myrtle), Townsend's, yellow, MacGillivray's, Canada, Wilson's, northern waterthrush, yellowthroat, or redstart. The most common and showiest is yellow warbler. This is the all yellow "Summer Yellowbird." I watched one building a nest on a cottonwood branch above our heads. She was using cottony stuff (probably from cottonwood fruit) and lacing it together with fine grass blades and cobwebs. Nothing came of this nest; she had not yet laid any eggs when a starling ran her off. On June 25, not far from that nest site, I watched two males fighting vigorously in the air, using beak, wings, and feet. It didn't last long; one gave way and left. We thought this meant there was a pair and an active nest nearby. In early July the adults are immensely busy carrying insects to fledgling young. Then all summer all members of the family are flitting about the bushes keeping the insects under control. They are always in view when we eat in the river-room (the picnic spot). I think humans, with all that naked warm skin pulsing with blood, are really tremendous insect traps. No wonder the warblers hang around. Any time we show up for lunch, here come the mosquitos. And so here come the warblers. By early September the mosquito season is over, so the warblers go.

The other regular summer resident warbler, Wilson's, is a yellow bird, too. Most casual lookers confuse the two. There are differences: Wilson's is a little smaller, he has a darker greenish yellow back, there are no breast stripes, and he has a small cap of shining black feathers. The place where you see each one is an important clue: yellow warblers are in the middle of shrubs or brushy trees, or higher; Wilson's are most often low, where bushes hang over the river. Except for this detail of habitat you couldn't even guess at the identity of the females, since neither one has a striped breast or a black cap; Wilson's warbler is just a little smaller and her yellow a little more toward green.

Audubon's warbler is the earliest to come and the latest to leave. Sometimes he stays around all summer, nests, and raises a family nearby. These are bright little butterflies of birdlings, blue gray with trimmings of black, white, and yellow. The yellow is throat and rump and little cap; the white is two eye-rings, two wing patches, and lower breast and belly; the black, a big flowing tie below the yellow throat. Audubon's warblers may come by mid-April and are still in the thickets in late October. Once I saw one in November. Peterson says they nest in conifers, but there are practically none in my study area: a young Douglas fir near the highway house, a dozen or so junipers scattered along the river, two or three small limber pines on the sagebrush benches. I wonder whether the few who summer here do find the few conifers? Or are they bachelors? Or are they aberrant or independent souls who don't believe that every Audubon's warbler must nest in a conifer? One year a male Audubon warbler and a male mountain bluebird struck up a friendship that lasted all through May. They were always near each other, hopping around in the sagebrush, sitting quietly on the top rail of the buck fence, fluttering off over the meadow. Bluebird would never try the thickets, so it had to be the Audubon who left his "proper" habitat. But when the time came for raising families, early June, Audubon disappeared, and bluebird devoted himself to the brood in the birdhouse.

The rest of my thicket birds are now well mixed, in the list if not in their bloodstreams. They are the grosbeaks, buntings, towhees, sparrows, juncos, blackbirds, orioles, tanagers, house sparrows, finches, and crossbills. When I started this study the listing was different, and I disagreed with it, but gave in to the experts. Now those experts have come up with the list above, and I'll have to give in again.

This group of birds can eat seeds, as most birds can't. Most of them are thicket birds, requiring nearby shrubs to duck into for shelter. Any time of year, some of this tribe will be around the riverside thickets: redpolls, tree sparrows, Harris's sparrows, pine grosbeak, evening grosbeak in winter. In spring and fall, lazuli buntings, green-tailed towhees, song sparrows, chipping sparrows, white-crowned, white-throated, Lincoln's sparrows, red-winged and Brewer's blackbirds, grackles, cowbirds, western tanagers, pine siskins, goldfinches. In summer, song sparrow, redwings, grackles, cowbirds, and Cassin's finches.

Black-headed grosbeaks come to our thickets every summer, and we see enough of them to feel we belong together. They are nearly as big as robins, and the males are almost as showy in their outfits of rust, yellow, black, and white. The head is black and the big bill white, and wings and tail are boldly black-and-white. The rest is yellowish to rusty, brightening to sunny yellow underparts, subdued on the back by dark stripes. The call note is a very sharp CHIP!, the song a sweet loud warble. The main characteris-

THICKETS

tic is confident calm. To see one pry off a piece of suet is to have the sensation of watching a slow-motion replay of a Harlem Globetrotter making a basket. I've seen them fight off chipmunks and house sparrows and never ruffle a feather or give a real glare. Calmest fights you ever saw.

A singleton is the indigo bunting. I'd known him in the forest preserves near Chicago, but it was forty years since I'd seen one. On May 21, 1971, here came one to the gooseberry bush near the feeder, bluebird blue all over except for black accents, somehow precisely filling an empty space waiting in my mind. The book says he is really rare in Wyoming, so I had better cherish the memory.

His cousins the lazuli buntings are in the river thickets as well as the fence rows.

Redpolls are occasional winter visitors to our thickets. They are perky little twittery sparrows with bright red caps set forward over their eyes, and the males have pink breasts. They nest in the far north in tundra scrub, so they don't want the real woods. They are in the thicket edges and the fence rows, where they seem to find plenty to eat. One December morning I saw one filling up on the seeds of a tumbleweed the size of a washtub, anchored by the wind against a fence. I've seen redpolls as late in spring as May 2 and as early in fall as August 20. The shortness of the summer between those dates makes me wonder if perhaps some nest on our Wyoming tundra above 10,000 feet. Or maybe these late and early ones are bachelors having fun and not concerned with nesting. If neither of these guesses are right, it means they are extraordinary little flyers who cram two 1200 mile flights, courtship, nest-building, egg-laying, brooding, hatching, and raising a family into as little as 110 days.

Fluttery little juncos are likely to fill the edges of the thickets in twitters in April and May, then in September and October. "Tut-tut!" they say, "tut-tut-tut!" They are never still. Juncos are ground birds, skittering along just above ground or scratching among the leaves. They build neat, soft, grassy nests among the woody cinquefoils and sagebrush just a little higher than my study area in the edges of the pine woods. I have no June, July, or August records from the study area, but many in April and May, and

September and October, and a few in November, December, and January. The little fellows have their problems. One October 30 I saw a northern shrike fly up from beside a birch to a dead branch of a cottonwood. He was carrying something quite long and pointed. He maneuvered it about, rather clumsily, trying to find a small fork of twigs that would hold it. Suddenly I recognized the object; it was the whole tail and part of the rear end of a junco, the dark inner and white outer tail feathers telling the pathetic story.

Tree sparrows are winter birds and not very common here. Like the redpolls they nest on the edges of the Arctic tundra and so just possibly on our high local tundra not many miles away. I see them most often from November to April on the ground under silverberries and willows back of the motel and in the lower pasture. Generally a soft winter lisp—a low inconspicuous "Tseet!"—tells me where to look. Busy little earth-scratchers, wasting no time talking, they're plain light gray below and on their faces, tweedy striped sparrow-brown and tan above. There are two good field marks: a red cap and a dark stickpin in the middle of the breast. In the intoxicating air of March they burst into song like the rest of us, a short enthusiastic little ditty that you might render "Whee! Chi-chi-wht!" One Christmas Count day I watched a delightful little company of eight tree sparrows on the Rocking Chair Ranch, scratching away in the brown needles and among the knotted brown roots of a big spruce.

In late April, perhaps in the midst of a wild swirling blizzard, perhaps under shower-patterned skies, comes a whistled call, repeated over and over. One year I wrote it down this way: "Oh-la-di-da-chleswh!" (four whistled notes ending in a trill). Another year I heard it this way: "Chee! Cheedle-cheedle-chee! EE-ee-ee!" Must be the same four whistles and a trill. I know it when I hear it, the call of the most elegant of the sparrows, the aristocratic looking white-crown. Follow his song, and you'll find him singing with power and enthusiasm from the tip of something—a willow spray, the transformer pole, a fence post, an insulator on a power line, the topmost twig of a tall cottonwood or of a lowly sagebrush. The height above ground doesn't seem to be critical; it just has to be the very top of whatever it is. The singer is big for a sparrow but not as big as his cousins the cardinal and evening grosbeak. Most conspicuous field mark is the brightly striped black and white head, looking big because the feathers seem to stand upright. The unmarked throat, breast, and belly are one evanescent pearly gray. The back is a trim, well-groomed pattern of black, brown, and tan; the legs and feet are pinkish brown. On a September morning one hammed up his elegance for me, looking self-consciously dramatic as he sat upright on the sign at our gate, catching the full early sunlight, with the rich dark red contrast of a wild rose bush behind him. I put my foot in my mouth one day when I described a "related species" to a birding friend. These similar birds appeared every fall, sometimes in a flock of their own, sometimes flocking with the white-crowns. They were as serenely distinguished as the white-crowns, only the striped heads were brown and tan, not black and white, and the breasts were beige, not gray. I was abashed when my friend laughed and told me they were the young of the year.

For all his style, the white-crown is plebeian in his tastes in food, and not shy about who sees him eating it. I've watched him pick over manure without losing his aplomb, attack the seeds of many species of grass with equal interest, and vary his diet with

high-protein trout flies on occasion. I've watched him bathe elegantly under the hose-spray on the lawn; and just as elegantly present seeds to a big demanding young cowbird. Once there was an early June snowstorm, weighing down with heavy wet wreaths the shrubs already in full leaf. I watched a white-crown, with elegant timing, gather a drink drop-by-drop from the dripping branches.

In May and September I see Lincoln's sparrows, inconspicuous little birds trying to be even less conspicuous than God made them. They will be in the willow brush near the bridge in the weed patch. Once in a while in late May an over-excited male will mount a weed stem and sing a short, bright, burbly song, something like "Sweetch-sweetch-gurgle-gurgle-oh," then duck and hide. I think the ones we see must nest in our own high mountains, up in the tundra on the Wind River peneplain, because they are here so late in spring and come back so early in fall. A little smaller than the white-crowns or the white-throats, Lincoln's is a very neat little striped bird. Fine head stripes are brown and gray. There's a broad buffy band across a light gray breast and very fine brown pin-striping down throat and breast. The pin stripes usually gather to form a small dark star in mid-breast.

Song sparrow, best-beloved of the thicket birds, looks something like Lincoln's sparrow. His markings are warm reddish brown on light gray; head is striped, back is tweedy, throat and breast are heavily striped, the stripes running together to make a dark star in the middle. He's a cherished regular resident for most of the year. His bright song announces the arrival of the first one in late February or March, and the last one may not be gone by Christmas. He is quite shy, never as confiding as the chickadees but nearly as curious. Often as I stalk a song sparrow for a good look, I find that he in turn is stalking *me* for a good look. He's a better people-watcher than I am a bird-watcher. Usually I hear song sparrow before I see him. He will be close to water and almost always in a birch or willow, usually singing from the end of a branch. Most of the time he sings with style and passion, but he may drool out the song with no affectation at all, just pausing a second from hopping about in the underbrush.

The song is easy to recognize, even though it has endless variations. I have a list of 26 different ones, and I've heard plenty more I haven't phoneticized. My mother taught me, back in Vermont, that song sparrow said, "Maids! Maids! Maids! Put on your tea-kettle-ettle-ettle!", but none of my Wyoming birds says just that. Nearest to it is one that goes, "What? What? What? oh-my-that-isn't-far-to-see!" I named him "Far-to-see," and I think that same bird nested in the big willow bush beside our river-room three years in succession. He had several variations:

"Chip! Chip! Isn't-far-to-see!"
"Chip! Chip! Chree-hard-to-see-through!"
"Tse! Tse! Tse! Not-far-to-see!"
"Chip-chip-chip-chip-chip-chip-its-far-to-see!"
"Chi—chi—chi—too-far-to-see!"

Some of the others go:

"Chip-chip-chip-chip! Whree! What? Tzip!"
"Ch-ch-ch-ch-come ON-swee-swee-swee-eee!"
"Hi-hi-hi-low-low-cheer!"
"Chee-chee-low-low-low-warble-trill!"
"Chir-chir-chee! Oh-all-you-want! Cheeeee!"
"Chip! Chip! Chee! What-you-say? Oh-wheeee!"
"Chip-chip! What? Here-to-see! That's right!"

When I read over those songs, I'm struck again with the muscle, energy, and bright personality in this little brown singer. I have read that the song expresses none of those things, that it's a purely instinctive expression of territorial rights. I go along with that to a degree. There are several pairs that I see regularly in summer, each pair with a territory maybe a hundred feet across. The pair just below the bridge, and the river-room pair, each claim a bit of both banks of the river. Then there's a pair by Mabbotts', another just below Leseberg Islands, and one in the swamp this side of the New Slough. They do "sing their territories," but in addition I am sure that the considerable variation in song connotes creative ability and showmanship. And I am sure that the personality differences are real and not something I mistakenly read into my observations.

Four species of blackbirds are familiar to me in the thickets: red-winged and Brewer's blackbirds, grackles, and cowbirds. These are all about the same size, and at first they all looked just black.

It took me a long while to learn to recognize each species at a distance, but after I knew what to look for it wasn't hard.

Redwing and Brewer's are middle-sized, a little smaller than robin. Identification is certain if you are near enough to see the red and yellow shoulders of redwing or the white eyes of Brewer's. But as they fly overhead you can still identify by the call note: Brewer's is a hard-edged "Tchk!", redwing's a juicy "Tchook!" Brewer's seldom sings in the thickets, redwing generally does. Redwing's song is accompanied by curious contortions: half-spread wings, coverts raised and spread separately like a second little top pair of red wings, head lowered and thrust forward. The song itself is a loud, half-whistled "Ka-kooeeeee!" or "Kong-dreeeeeee!" These two species are thicket birds only in early spring, when the cold wintry winds move them to find shelter before nesting time. As the weather warms they leave the thickets to forage in the fields. When it comes time to nest, redwings head for the cattail patches of the Motel Slough, Brewer's for the gooseberry bushes and sagebrush of the fence rows and ditchbanks.

The common grackle is the largest of the four blackbirds, about three inches longer than redwing or Brewer's. Grackles are new to my study area. The first one showed up, casing the joint, in mid-June 1967, and stayed for a week. The first resident pair came in 1969; another pair, or the same ones back, came in 1970 and again in '71 and '72. Grackle has a long tail, wedge-shaped, a fine field mark, as far away as you can see him. Nearby, that long tail, his cocky strut, bright golden eye, and the blue shine on his head are not to be mistaken. His so-called song is a whistled squeal, rather like an unoiled wheelbarrow wheel. He is only accidentally a bird of the thickets, belonging more to the open river-woods.

The real thicket blackbird is the cowbird. He's the smallest of the blackbirds, an inch or so shorter than redwing. When you see the male close by, you can see his dark brown head, dark eye, and the bluish shine on his body, but at a little distance he is black all over; so the surest field mark is his thickish bill, much like a sparrow's, where the bills of the other three are narrow and sharp-pointed. Cowbird has quite a song, somewhat wheezy like the others but longer, more varied and bubbly. Every year there are a couple of pairs of cowbirds in my area; one pair has a stake-out near the bridge, one pair downstream from our "river-room." The reason they are thicket-birds is the unpleasant home-making (or home-wrecking) habit of the female, who believes in Women's Lib for sure. She does not like housework or raising kids, so she has worked out a system for avoiding both. She lays her eggs in other birds' nests, like yellow warblers or white-crowned sparrows. The hatchling is so much bigger than the tiny warbler or sparrow babies that he generally either crushes them or forces them out of the nest, so the frantic little stepmother has only him to raise. She usually does try to raise him, working hard to catch enough insects to match his monstrous growth. On balance, cowbird's selfish habit seems not too successful; it keeps the species going, but not increasing—around here anyway. Of course life is much easier for the irresponsible cowbirds than for all the hardworking feathered homebodies, their neighbors. During the busy nesting season, the cowbirds give the impression of light-hearted partying. They are often seen walking about among the horses or cows, even up and down their backs, supposedly looking for insects but apparently enjoying looking at each other more, and they spend considerable time love-making in the thickets.

Pine siskins I expect in late April or early May, though in 1965 the first one came March 31. They nest not far away, close to the lower edge of the forest. You see them in the pine trees, usually not till you hear them first—gentle, conversational hisses. "Sisk?" asks one, and "Sisk!" answers another. That's why they're called siskins, I guess. They get back to our thickets in late August and September. Now and again I see one or two for days at a time at my feeder. I have one January record of a flock of a dozen back of the motel. Siskins are the plainest of plain brown sparrows; you have to look awfully close to catch the bit of yellow on the bases of wing and tail feathers, visible only when they fly. It's hard to believe they are close relatives of the goldfinches. Often they flock together, making you think of plain country cousins with their brilliant city slicker cousins.

Oh, the goldfinches! Yellow as candle flames. Black cap and wings. Too brilliant to be believed. They are not nearly as common as those plain cousins the pine siskins, but I do see them every year. They fly like roller-coasters, in up-and-down golden swoops, as they look around for dandelions gone to seed. Some must nest nearby, for I see them once in a while in June, July, and August. On a July 17, surely in the middle of nesting season, a male and female dropped down from willow brush to wet black sand at the edge of the river and drank deeply, each motion reflected in both sand and water. It was the very same day, and not far away, that I was watching a great horned owl taking a siesta on a cottonwood branch fifteen feet overhead. With an explosive flash a goldfinch sailed up to a dead stub of birch nearby. Alertly he ducked into the middle of the birch, then kept up what sounded like a sally of teasing remarks. Maybe he got the big fellow's goat. The owl first fidgeted, then flew across the river.

Once in a great while in winter or early spring come the beautiful evening grosbeaks to the riverside bushes. They nest in the evergreen forests of Canada, and I expect them only as off-season tourists. It was a pleasant shock to have a pair turn up one Memorial Day and a joy to see them stay around a while. Generally it's severe weather when I see them, but the wandering groups seem carefree. I've seen as many as twenty-eight together. They don't lack for food—there are always silverberries, rosehips, gooseberries, cedarberries—so they are as confident as their cousins the black-headed grosbeaks of the summer season. Like them they are big and slow moving, and their coloring is not too different. The male is yellow, black, and white: big white beak and big white wing-patches, black cap, tail, and primary feathers; and all the rest yellow, very bright on forehead and eyebrow-line, very dark on the face, lightening gradually to bright yellow again on rump and belly.

The Woods

FORTY ACRES OR SO OF MY STUDY AREA are woods and thickets in four distinct patches. From west to east (left to right on the map) or downstream from the bridge, they are the Home Woods, Back of the Motel, the Slough Woods, and the Lower Pasture. Less than half the area is real woods; the rest is thickets, and that's where most of the action is. But some creatures prefer the quiet and isolation of the woods, and they must have their own small chapter.

The trees of the woods are narrow-leafed cottonwoods. Their well-spaced trunks rise dark and grotesquely crooked, lifting the canopy of slender green leaves thirty or forty feet above the bark-littered ground. There is a different feeling here, a slower rhythm than in the thickets and a sense of space. The cottonwoods have a corky bark that climbs the trunk in a maze-like pattern of ridges two or three inches thick and not very tightly attached. Long lumpy stringers of bark fall off and hang from limbs they've been caught by on the way down. Where the canopy is closed the floor is fairly open and predominantly brown, littered with bark and small gray branches, and with the mouldering corpse of a fallen tree here and there. Where there are holes in the ceiling patches of thicket lie below—wild roses and gooseberries, young cottonwoods, willows, and silverberries.

All of the woodland is level. It is the climax of the growth that began with the bare gravel bars like those that show up raw and new after the spring flood. The woods require a lot of ground water, though not as much as the swamps. It is found in abandoned river channels and on old islands, where it is only four or five feet below the surface.

You don't need any special technique to traverse these woodlands; you simply stroll along easy pathways. These are pleasant places, but curiously empty. Most of the insects, mammals, and birds seem just to cross the woods, not to live here.

In the Slough Woods are a family or two of flying squirrels. I suppose the openness is safer for their gliding than the thickets are, and they need the flicker holes for nurseries and bedrooms. One of the pine squirrels lives in this patch of woods. I have met deer and moose on their way across it, too. The main residents are birds. Some of them prefer the woods for one reason or another.

Several times I've seen goshawks in the big trees. These are the big blue hen-hawks, among the fiercest birds there are. I suppose from the name that they will even take on geese, and geese are some challenge. Ordinarily they nest higher up in the mountain forests, and I see them only as they pass through. I have great respect for them as predators. Once when Joe was logging in the Crooked Creek drainage, he watched a goshawk kill a big snowshoe rabbit. The animal was too large for it to lift and fly with; yet because of Joe's presence the hawk seemed to feel that he must carry it away. So he skidded it, flying slowly only inches above the snow, so that the snow bore most of the rabbit's weight. Once, when we lived at Lava Creek and I was a new bride fresh out of Chicago, I heard a terrified scream from my one hen and rushed out the kitchen door. A neighbor had lent me the hen to brood and hatch a clutch of eggs to start a flock. Now she was crushing herself into the scant shelter of a sagebrush as a goshawk attacked with great ferocity. One look printed indelibly on my mind the powerful blue-gray body, the flame-red eyes under a black cap, the wicked hooked beak, and the yellow legs stretched straight forward, reaching with curved black claws for the chicken. I had a dishtowel in my hand, and I used it for a whip. That goshawk stayed for a real flogging that tore the towel to slivers before it gave up and beat a retreat and let me tend to a hen that was by then deep in shock from sheer terror.

On the first of March I found a dead goshawk lying on fresh snow in a willow thicket under cottonwoods. The snow had fallen less than three days before. In that short time the bird's body had been opened up from the top and every shred of meat taken off the bones. The skull had been beaten open, too, and the brains were gone. The only tracks around were of magpies. The gutted and empty skin and bones still looked tenacious and fierce. I took the carcass home to use as a model for a painting, trying to reconstruct the scene as I found it, with snow, willow stems, magpie tracks, and eventually the magpie itself, looking vainly for a last bite. I called the painting "Warrior's End."

A pair of bald eagles has been patrolling this stretch of river for many winters. I don't know where they go in summer. They are by no means woodland birds and are found in this space only because they need spy trees at intervals for resting and looking around. For years there was an eagle tree at the lower edge of the Slough Woods—a very big, long dead cottonwood that stabbed a gray finger toward the sky. If either eagle was in the area and not soaring about, it would be there, a huge dark lump on top of the gray finger, the white of head and tail catching the morning light. Two or three years ago in a tremendous windstorm the eagle tree crashed, its shattered body spanning the river, making a dam. The rancher fastened a cable to it and worried the top upstream with his tractor, so the big carcass lay parallel with the river on the Killdeer Bar. Since then the eagles have experimented without finding another such lookout spot. There are four trees to be checked for eagles now, one in each of the four woods areas.

When we first built a cabin here and moved in, great horned owls nested in the Home Woods on top of a great tangle of branches and twigs put together by magpies. We had a small shepherd pup named Whiskers, black except for white feet and a white mark like a comma on the top of his nose. He went into convulsions one day. It was soon over, but we thought we should take him next day to the vet (eighty miles away). By morning he had disappeared. Our grown shepherd, his older brother Monkey, was much distressed and spent hours in a thorough search, trying out all the bushes, culverts, slab piles, and boxes. Our best guess was that he'd had a convulsion on the lawn in the moonlight and had been seen and picked up by a great horned owl. The time was April, when there should have been young in the nest. A few weeks later we had a confirmation of sorts. I was hanging up clothes on a line strung in the woods, away from the punishing winds in the open. Monkey, exploring, stopped abruptly to inves-

tigate something below the owl nest. He picked it up gently in his mouth and brought it to me, nudging my leg and delivering the object he was holding so delicately into my hand. It was the skull of a small dog, completely cleaned except for the nostrils and a patch of skin on top of its nose. The skin carried black and white hair, arranged in a pattern like a comma.

Great horned owls are among the biggest and fiercest of their kind; they stand knee-high to a person and boast a five-foot wingspread. Their soft feathers are mottled and barred brown and tan, with a white throat patch and three-inch-high "horns" of brown feathers. Their fierce yellow eyes glare between black lids and black pupils. The downy silence of their flight is in contrast to the neck-pricking hollow gruff sounds of their calls. Sometimes they bark sharply, like a dog, and sometimes two owls, well separated, will carry on long conversations in a choppy rhythm of hoots and barks. We hear them calling and talking once in a while in the night from September to April. Strangely, all my sightings are from April to September. They talk in the winter and display themselves in the summer.

On a July morning a commotion of robin voices above our neighbor's house brought me over in time to see a great horned owl swoop out of the branches and down low across the woods, pursued by a pack of screaming robins—quite a switch.

The Slough Woods are the owl woods. The one surviving heron nest platform has been taken over by great horned owls, and they generally raise a couple of babies there each spring. They are not particularly shy or cautious, and I often see one or more of them. One May 29 I watched one adult and two half-sized youngsters with feathers scattered over fuzzy whitish down, all on the nest platform. On June 19, in the same place, the adult flew toward me and kept performing clumsy aerobatics between me and the nest, successfully keeping me from taking any pictures of the big youngsters. The next Monday there were two big young owls (looking like the adults except that the spots and bars on the feathers were blue-gray instead of brown and tan) sitting side by side on a high branch not far from the nest. They were in the same place every Monday morning from that time into August. In the same woods that August, one adult let me get as close as fifteen feet without stirring, and another adult sat on top of a brush pile at about my eye level and let me walk all the way around it. First its head followed me around with a steadily twisting neck, slitted yellow eyes gleaming; then in a gesture of either trust or contempt it snapped its eyes away from me entirely and seemed to look intently at something in the opposite direction.

August 22, half a mile south of the nest platform, high on the Jakey's Fork moraine, I found a great horned owl moving strangely on the ground among the sagebrushes and the granite boulders near the place where the REA line crosses the hill. When I came close, he threw himself on his back, feet in the air, needle-sharp talons widespread, two toes forward and two back, quivering legs completely covered with grey feathers to the very talons. His yellow eyes flamed; his beak set up a clicking like drumsticks. He was a youngster, I could see, with extra soft plumage and steel-blue coloring. I was sure he must be injured. Since I was wearing leather gloves, I was brave enough to reach in with both hands and grip his feet, lifting him into my arms. The answering grip of his feet hurt my fingers, even though the talons didn't pierce the leather. With his feet near my waist, his unflinching fierce eyes were level with mine. It took most of the walk home to transfer one of his feet to the one hand, so I could use the other to hold his wings folded and close to his body.

At home, Joe and I examined him and found that his left humerus was broken. He had probably flown into the REA line. The break was close to the shoulder. We couldn't see how we could set it but thought that before we gave up we should ask some of our friends who might know more about it. Joe made a firm leather splice around the owl's leg and fastened the other end of the thong to the leg of a ladder at a shaded corner of the outside log wall of the house. We left then, thinking he needed solitude for awhile. When we went to check him later, he was gone. He had carefully and cleanly unspliced Joe's tie. (Joe was affronted, because he is proud of his knots.) We searched for him a long time among the sagebrush nearby, but without any luck. We could not see how he could survive without the power of flight. The only chance would be to catch mice. Well, maybe he could make it afoot. Mice were plentiful. He really might be intelligent enough, quiet enough, to do it. The next we knew of the young owl was 12 days later, September 4, in the woods back of the motel. The ground was littered with blue-gray owl feathers over an area at least 100 feet across. The only possible answer was that the young owl with the broken wing had *walked* a quarter of a mile (half the distance to his home nest and in exactly the right direction) and at that point had been attacked by some determined predator. He must have sold his life at a very high price. The predator surely had a hard fight of it. Who could it have been? Coyote? Mink? Otter? Badger? Bobcat? I found no evidence to name the attacker. The very next day I saw the young owl's brother attacked by a flicker and six magpies; they forced him to leave that piece of woods. I'm sure the assailant of the night before was much stronger and fiercer than a flicker and six magpies, but I cannot help wondering if word of that battle and its ending gave new courage to certain birds.

Four species of woodpeckers live in the woods I watch; red-shafted flickers, sapsuckers, hairies, and downies. By definition they belong to the woods. Even the flicker, with his wide dietary tastes, has to make his home here. The others not only live and raise their families here but find all their food here, too. Very specialized birds they are, their whole anatomy bent toward getting nourishment out of vertical containers with tough rigid walls: trees, that is. Imagine spending your whole life beating your head against the walls! Woodpeckers' adaptations to this demanding routine include a short heavy muscular neck, a thick skull, a beak shaped like a chisel with cutting edge held up-and-down, and tail feathers with barbed shafts strong enough to sit on and use for a working platform. The small expressionless woodpecker eye, compared with the large liquid eye of a nighthawk, makes you think of a camera rather than a device for communicating emotion. A hedge of stiff feathers protects it from wood dust. The ears are large, though invisible behind stiff openwork feather protectors. The feet have two toes together facing forward and two wide apart facing backward for more efficient propping.

Armed with all this array of technical equipment, a woodpecker flies to a suitable dead tree, lights against the bark in a vertical position, propped by the specialized feet and tail, and listens. He hitches around in short jumps, prospecting and listening over a good-sized area. When his mind is made up he sets to work industriously, using the hammer that is his head and the chisel on the front of his face, and cuts a hole into the wood until in a short while he brings out a grub and eats it. (I wonder if this woodpecker habit is the source of the word "grub" for human food.) This is surely a hard way to make a living. I wonder if the woodpecker's glassy-eyed stare indicates a state of constant concussion.

For a home, the pair of woodpeckers excavate a hole in a dead tree in the same manner as getting out a grub, but they put in much more time at the same spot. They drill out a horizontal passage big enough to crawl into, then a bigger vertical shaft down from it. At the bottom the eggs are laid on a bed of chips with a few feathers. Suitable dead trees are much sought after for residences, and not only by woodpeckers. I know one or two in my study area that are apartment houses, with a dozen or so nest holes, one above another. Some may be vacant in any given season, but several are likely to be occupied. Woodpeckers generally use a house for just one nesting season, making another next year. Abandoned holes are likely to be snapped up by new tenants like white-footed mice, pine squirrels, flying squirrels, saw-whet owls, tree and violent-green swallows, chickadees, nuthatches, house wrens, starlings, or house sparrows.

Downies are the smallest of our woodpeckers, about the size of house sparrows. (Once I saw one repeatedly challenge a house sparrow to a fight, attacking him with quivering wings and open beak. The sparrow just stood his ground for six sallies, then flew away. I gathered that the house sparrow is considerably calmer of temperament than the downy). They are black and white, wide-striped across face and head, plain white back and belly, black wings and tail spotted with white. Males have that red spot on the back of the head.

Hairy woodpecker is downy's big look-alike. Big as a robin, with head bigger in proportion than downy's, the black, white, and red patterns match, feather for feather.

Both of these are year-rounders. The dead trees yield food in winter as in summer. The tap-tap of the downy and the rap-rap of the hairy are regular, expected sounds of the winter woods, more than in summer. One reason, I guess, for more tappings in winter, is that with a constant supply of potential food, the greater food needs of cold weather simply mean more work. Another reason is that in winter the sound-dampening barriers of millions of leaves are not there.

The sapsuckers are here only for summer. Late comers (my earliest date for first arrival is April 20 and the latest, June 6), they are early to leave, too, seldom staying into September. "Yellow-bellied sapsuckers" the books call them, but that's an inept name. The belly isn't yellow at all, or not more than a touch of cream on white. The main field mark is a long white patch shaped like a dagger on each wing, visible whether in flight or at rest. How about "dagger-winged sapsucker" instead of yellow-bellied? The head of the adult is brilliant in a bright, contrasting pattern of black, white, and red stripes. As field marks, though, the stripes are too narrow to show up; instead of calling attention, they break up the head shape and make the bird less instead of more visible.

THE WOODS

The thing that make sapsucker's season so short is his specialized eating habits. His race has learned to like to drink sap. They also use the sap as bait for insects. So instead of listening for grubs, the sapsucker looks for healthy young smooth bark. He drills small holes (only into the sap-bearing layer just under the bark), many holes, in neat rows clear around the tree or branch. The sap wells up into them and often runs over. The sapsucker watches, and at the proper time comes back to drink the sap and eat the insects that like to drink the same sap. So their time in my study area is the short part of the year when the sap is rising.

The largest woodpecker in my study area is the flicker, generally the red-shafted, bigger than a robin. The pileated, much larger, is listed for Wyoming, but I've not seen him. He is not chained to the woods like the others, because much of his diet is ants; but he likes the tall trees and needs them as nesting places. He is a loud, conspicuous character, and a welcome early sign of spring. One year he came January 30; usually it's early in March. There'll be loud uninhibited drumming, shrill loud repeated calls, conversational "whicker, whicker, whicker," much show of pretty feathers. They like to watch people and seem to like to be watched: they certainly seem to ham it up, with displays of spread tail and wings. Before the leaves are on the trees, it's a welcome sight to see the big birds fly past, swoop up to near the top of a tree, land against the trunk, cry out loudly (to get you to look?) and start grooming the bright red underside of wings.

I wish I knew more about their family life. I know they are cheerful, loud, and competent whittlers, carving out ample nest holes in dead trees or with equal enthusiasm in the gable ends of cabins. On a May 5 we watched flickers mating on a cottonwood branch above our river-house. On August 5 a big juvenile sat on a suet chunk on our feed tree and yelled loudly for the folks to come feed him. His folks had started to wean him, so they held out quite a while before they gave in and pried off suet beside his feet and crammed it down his craw. By August 10 the weaning was over. The big baby was doing his own prying off and stuffing down.

The only flycatcher who is a regular resident in our woods is the western wood pewee: the eastern and western kingbirds belong in the open; Say's phoebe is a desert bird; all the Empidonax tribe live in thickets, each species in his special part. For years before I ever saw the pewee I knew him by his voice, a sad, resonant single note high in the green canopy: I phoneticized the sound as "Prweee-p" (a little different from the "pee-ur" in the bird books' description). I called him "cottonwood ghost." I would first hear him in late May or early June, then every day all summer; the lonesome call would be gone one day in early September. I identified him in 1963; somehow, after that, I had little trouble finding him whenever I heard him call. Strange how much it helps to know where to look and to have an image ready-made in your brain, waiting for your eye to locate and possess. Pewee is the size of a small sparrow, olive gray with dark breast and flanks, whitish throat and belly, and two white wing bars. He sits motionless on a dead twig, then suddenly rises a few feet into the air, snaps his beak, and returns to his perch with a small insect. There are no wasted fussy motions.

Tree swallows and violet-greens nest in the woods but do their insect hunting elsewhere, over the streams, the sloughs, and the hayfields. Both species make use of woodpecker holes. They usually wait until the holes are a year old, because then the holes are vacant and for rent, but one year we saw a real fight as a pair of tree swallows tried to drive a pair of downies away from a nest hole they had just finished. The downies repelled the invaders, but it was a real battle while it lasted, fought in the air with wings and feet. Neither species is equipped for armed aerial warfare, but both put on a good show. The year-round downies have the countryman's stubborn attachment to his territory; the summer resident swallows are the "dudes" of this dude ranch country; so it seemed appropriate that the year-rounders should win.

Violet-greens, unlike tree swallows, are not tied to the woods. They make their homes in all sorts of crannies. I find them in niches in the badlands cliffs; a family nested in a crack in the back of a board under the eaves of the wellhouse in Dubois City Park; another at the root beer stand. But both species do live in our woods and raise families every summer. In late July for ten days or so the whole woods quivers with their twittering. They are responsible parents: on August 1 one year I watched a determined vigilante committee of eight or ten tree and violet-greens together chase a pigeon hawk out of the woods and across the river. This was incredible—the hawk is much larger, and swallows are all but defenseless. It shows it's the spirit that counts.

When the youngsters are airborne they leave the shelter of the woods and gather on dead twigs in the thickets and on wires in the open. I see adults swooping past wires in early August, dropping insects into open baby mouths as they go by. One August 5 I saw two adult tree swallows on a wire, one at each end of a line of eleven juveniles! Once on August 7 I saw thirty-six youngsters sitting on the tiny top twigs of the crown of a dead cottonwood or playing in the air nearby. It was a hot-rod time. They would fly drag races all over an airy track with imaginary boundaries, turning tight circles, flying straight toward each other, playing chicken. Suddenly one or two would drop out to a twig and doze awhile in the sunshine. In the evenings in this part of the summer there may be up to fifty gayly cavorting over the river; I surmise these are several families of both species all having fun together. It won't be long till the wires are crowded every day, then by the end of August all the swallows will be gone.

The noisy crow-and-jay family, the Corvidae, have four species for my list for the woods, but they are all visitors and not residents in this habitat. They are magpie, raven, Clark's nutcracker, and gray jay. Magpies nest in the thickets and use the woods only in passing. Lots of them live in my study area—my guess is at least ten families—so there is a considerable amount of passing. If there is any interesting activity in the woods, or anything dead, magpies will be along to investigate. They will discuss whatever-it-is with each other, light near the tree tops, and drop easily down the long stair-steps of branches to the ground. Bright, brash, and showy, they will strut across the open spaces and fly only when pressed. The black, dour ravens light often high on the crowns of the trees or on dead stubs. They will mutter darkly to themselves or to each other, hunching shoulders and looking around like conspirators. Something will attract the fellows this far every few days as they follow their skyway between the Dubois dump and Jakey's Fork, but I've never seen a raven deep inside the woods.

Clark's nutcrackers and gray jays are tourists. They live higher, in the pine timber above 7500 feet. But almost any time in late summer and early fall (family cares are over, there's plenty of food around, and the weather is fine) they may come around just to see the sights and how the flatlanders are doing. Since both species live in the forests, our cottonwood bottoms are natural wayside rest stops. They may come alone or in a family group. One time a nutcracker and a gray jay came together. The two are often confused, even by people who see them often: both are nicknamed whisky jacks, moose birds, and camp robbers; both are gray and white; and both are jays. But there are clear differences: gray jay has no black except eyes and feet; nutcracker has black wings with a white patch and a black tail with wide white edges. Gray jay has a soft purr of a voice; nutcracker has a loud raucous squawk. Gray jay has a short beak; nutcracker's is bold, big, and decisive like a crow's. In their own pine timber, the gray jays live low on the trees, the Clark's nutcrackers high in the tops. Gray jays are much more likely to come to camp and take snacks from your fingers. As tourists in the cottonwoods, gray jays are quieter and nutcrackers louder and more assertive. The odd couple who traveled together showed this clearly—the gray jay followed the lead of the nutcracker, who did all the talking.

Black-capped chickadees are found in the woods, but less often than in the thickets. They flit about for insect food all over the cottonwood trunks, conversing with each other most of the time. They live in hollows or niches, frequently in woodpecker holes.

The nuthatches are real spirits of the woods; I've seldom seen them away from it. Their tiny tooting says, "Be at peace. Everything's all right." The red-breasted nuthatch comes here, but the white-breasted really lives here. In fact, he lives so close to our river-house that I'll save details for the yards and gardens chapter.

Brown creepers, moving up the trunks like little mice, belong in the woods. They are scarce in my study area, maybe one in migration every three or four years.

House wrens and robins are woods birds; details are coming in the yards and gardens chapter.

Hermit, Swainson's thrush, and veery come through our woods in migration, though only hermit is strictly a woods bird. These three look-alikes—quiet, shy, trotting over the dead leaves and

THE WOODS

hiding behind twigs and brush—slip through the woods in mid-May on their way to higher ground. All three are brown above and buffy gray below with spotted breasts, shaped like robins but smaller. You look for the rust-colored tail of hermit, the buff face of Swainson's, the dark face and tiny breast-spots of veery. The hermit nests in the high, misty evergreen forests. I hear his quiet slow echoing song in evenings and mornings when I'm back-packing near timberline. According to the books, the other two might nest right here in our willows, but I've never seen either veery or Swainson's in summer. They must go higher.

Once in a while I see a shrike, northern in winter, the smaller loggerhead in summer. They use the very tips of the cottonwoods as lookout posts, but most of their activity is in the open.

A true bird of the woods is the warbling vireo. During June and July his voice is the voice of the treetops. Each of my four patches of woods has a summer resident pair. Vireo's slow conversational warble, often not consciously heard, is an intimate part of remembered warm summer days: "What do you know? Was *that* how it was? Do tell! You surprise me! Oh, I don't *think* so! Well maybe." He never really *says* anything but makes all the proper remarks to keep some unheard conversation going along. This is a very inconspicuous little bird, not as long as your hand, and sized, shaped, and colored to match the cottonwood leaves where he works on his insecticide job. His back is olive, his breast pale yellowish gray, and there's an indistinct light eyebrow line. His style is a slow flitting, matching the rustle pattern of the leaves in the breeze. You would never see him except for his curiosity: he is a people-watcher, at least sometimes. I have aimed my field glasses toward what I thought was the area of the treetop conversation, failed to locate the vireo, tried the naked eye again, and spotted him descending the long twiggy staircase toward me. Twenty feet is close enough. He takes a long hard look, then quickly becomes a rustling leaf and disappears.

There are three bright-colored birds of the treetops that I can easily confuse in the shifting lights and shadows of their chosen home. The black-headed grosbeak I have mentioned as a thicket-bird, but he is one that crosses over into the woods; the other two are Bullock's oriole and western tanager. They are all the same size (bigger than a sparrow and smaller than a robin), the males are gorgeous orange and black, the females nondescript olive green or olive brown. To separate them, the first test is by season. Western tanager is a spring-and-fall bird. He sings his loud, husky, robin-like warble (like a robin with a cold) when the trees are just coming into full leaf; then he goes higher up the mountainsides to the lower edge of the evergreen timber to build his nest, raise his family, and spend the summer. If you can get a good look, he is the one with the red head above the yellow-orange body and black wings and tail.

Bullock's oriole and the black-headed grosbeak are both summer residents. In my woods the oriole is the scarce one, the grosbeak the one you expect. The oriole is the most brilliant of all three, the main color flaming orange accented with black bill and bib, eyestripe, cap, and upper back; black wings with a big white patch; black in the center of his tail. The song is striking and pretty, a series of double notes. The few times I've seen him in our woods, I've heard him first, then been startled by the color, and wondered why seeing didn't come first. The only nest I have seen was behind the motel on June 5, 1965. It was a pendant cup, not as deep as a Baltimore oriole's, hanging from a cottonwood branch only about fifteen feet from the ground. Mr. Oriole was singing lustily from the very top of the same big tree. My earliest date for oriole is May 12, the latest, June 19. My only personal knowledge of oriole's diet comes from the 17th of May, when I watched one near my clothesline slowly eating silverberries, rolling each one around in his narrow black beak, then swallowing it whole.

The orange-and-black bird you expect to see is black-headed grosbeak. The species is more abundant, his season is longest (late May to mid-September), you find him both near eye-level in thickets and high up in trees, and he is quite regular about returning to last year's haunts. The predominating orange pattern is more subdued than either of the others, and it gives way below to bright yellow underparts. Like oriole's, the black back, and the sure field mark is the very thick, pale bill. This is the calmest bird I know, unless it's his winter cousin the evening grosbeak. To most birds you could attach the adjective "fluttery," but not to either of these. His movements are slow and thoughtful. He has a way of squatting on a twig and looking you solemnly in the eye. He inspects the food supplies at the feeder with care, then deliberately pinches off a bit of suet, cracks a sunflower seed, or takes a long drink of water. His song, along with western tanager's, is much like robin's—a regularly phrased and repeated warble—but grosbeak's has much more richness, tone variety, and resonance than either of the others. I've heard two other notes: a soft whistle (Wheeo-oo!) and a very loud, sharp CHIP! The youngsters seem slow to develop; perhaps this is a part of that deliberate nature. I saw a father feeding a big juvenile on August 23, only about three weeks before the family left for the winter.

The females of this group of three are harder to tell apart than their mates. Easiest is the black-headed grosbeak: she has tweedy brown stripes on a light olive ground and that enormous pale beak. The other two are both brownish green above, olive yellow below, with light wing bars. You look for the bills: Mrs. Oriole's is sharp and black, Mrs. Tanager's light yellow with a hump on the top side.

Just one of the sparrows must be listed as a tree-top bird as well as a thicketer. The white-crowned sparrow is the one, because he loves to sing from the very top of something, and that something may be a tall cottonwood, a long way up from the thicket where his nest is.

Houses, Yards, and Gardens

MAN, THE CHILD OF HIS HISTORY AND HIS NATURE, inevitably tends to gather round himself those objects that repeat his history and his basic nature. When it comes to living space, he almost everywhere tries to create close around himself a forest and a thicket. A house is modelled in a way after a thicket, with sheltering walls modified to let the breezes through, with burrows and runways, bedroom nests, places for food storage, places for waste disposal. Ask any pack rat about the validity of this comparison: he can find his way around a house as easily as around underbrush. Humans try to recreate the feel and look of the thickets by bringing cut flowers and potted plants inside and by planting grass, shrubs, and trees outside. So it is wherever people dwell, and of course it is so on the mile-across disk I live on. There are twenty families: two are new to this spot this year, and their thickets are still to come; the rest have either planted thickets, built next to thickets, or both.

Without sowing lawn grass seed, one can make a pretty fair lawn by just leveling the land and watering, then mowing what comes up. The yellow sweet clover, so rank and tough if left to itself, becomes a fine soft ground cover when kept mowed. The common wild grasses, brome and slender wheat, give that cover the body and look of a civilized lawn.

Fringed sage or cushion sage is a common relative of the better known big sage. In conditions of full sun, it makes a beautiful silver ground cover when kept mowed. If not mowed, it is a cushion plant, the needle-shaped soft gray leaves growing in a low dome out of which rise many foot-high fruiting stalks in late July and early August. These carry small yellow blossoms, brighter colored than big sage flowers and several weeks earlier. It wants to nose its way into a green lawn, where it's called a weed because of the gray spots it makes. It gets pulled up and killed. But where it is welcomed, transplanted, and kept mowed, it makes a smooth silver cushion, something like a moonlight-colored shag rug. Two particular advantages it has over grass are that it requires less water and needs mowing less often.

Gooseberries, wild roses, and silverberries, the ingredients of the wild shrubbery, are at home in people's yards, too, yielding beauty with minimum care. One of our neighbors is using big sage as the background of her garden arrangements and making little bedding plantings around the sage bushes. In her case, she had to guard against too much water, or her background shrubs will die.

Birches are hard to raise. They grow by choice with their feet in the water. It's hard to keep them well enough watered in one's artificial thicket. Cottonwoods need lots of water, too, but they have the faculty of being able to send long white tender roots deep into the desert ground searching for it. Cottonwood trees, if well-watered till mature, can live for years beside deserted homestead buildings. Of course, by the same fact, cottonwood roots are a great cause of drain problems in country plumbing. Once a well of ours had to be cleaned. It turned out that several bushels of fine, sleek white roots had to be cut off and put in tubs to be lifted out and thrown away. Roots like those will feel their way along sewer pipes and enter them by the little holes in the pipes of the drain field. Finally they fill them completely. "The drain field failed," we say. The backhoe must come and dig it all up and a full set of new drains laid.

But what shall one do? Cottonwood is the only tree that grows readily in our desert. Shall we then go treeless? Since we have man's instinct for trees, no. We'll just support the people who have backhoes and who import drain pipes.

Of course we have exotics. Every housewife who ever lived here has tried something to make her personal thicket more convincing. Most of the plants died, but some of them grew. It takes a plant of a certain essential heroism to brave an annual rainfall of about seven inches and a growing season of ninety days more or

HOUSES, YARDS, AND GARDENS

less; and certainly needed is a tender loving human heart and brain to foster the heroism.

Among the exotics that succeed are two vines. Hop vines are enthusiastic. They have perennial roots but annual tops. It's always fun to watch the long red fingers crawl out of bare ground in early May, reach out for last year's strings or wires, and unfold handfuls of green leaves. A friend in Casper gave me my first shoot of Virginia creeper, already rooted in a pot. "It will be slow to start," she warned me, and it's a good thing she warned me. It grew a quarter-inch the second year. I expected a quarter of an inch the third year and got six feet. The fourth year it was a thick canopy over the front door and marching toward a wall-cover.

Four non-native evergreens have been tried and may succeed. A Douglas fir brought down from a freshly timbered area at about 8500 feet, a little fellow a foot high, took hold, and ten years later reached six feet. Several Colorado blue spruces, brought in by a neighbor from a nursery, look good a year later; but I have a nursery blue spruce I set out as a two-year-old five years ago—the only survivor of ten set out at the same time and now only fifteen inches tall. Andorra juniper is a low-growing cedar in the yard of a friend in the mountains of New Mexico. She gave me a lot of cuttings last year, saying they'd come up fast. They barely weathered their first winter and a year later show only a little growth. Fitzer junipers, the common dwarf dooryard decorations at lower altitudes, do continue to live and look charming here, though even more dwarf than they are bred to be.

Upstream from us the ranchers have some beautiful golden willows, which are real trees though small ones. Only sixty miles away and 2000 feet lower, on the Wind River Indian Reservation, golden willows make spectacular windbreaks thirty feet high. So far no one in my study area has found them doing that well.

Exotic shrubs that do well with proper care include yellow roses, black currants, and lilacs. Red osier dogwood and chokecherry are exotics only to my study area: they grow as native both upstream and downstream, and they grow well as transplants right here.

In full sunlight delphiniums are marvelous. They will grow five feet tall if they have full sunlight and a wall to the north and will carry eighteen-inch spikes of blue flowers. Old-fashioned pinks, phlox, and sweet Williams make masses of bloom. Columbines, marigolds, oriental poppies, petunias, and pansies can be encouraged to respond. Sweet peas are fine but don't start blooming till mid-August.

Kentucky blue grass and orchard grass will enrich and make a velvety native lawn.

What mammals you find inside the houses, besides man and his pets, are likely to be mice, and most likely the white-footed mice or deer mice. I've seen a house mouse, the gray one with the long naked tail only once; must have been a fugitive from a moving van. People are jealous and arrogant creatures. Our objections to mice are based on their taste in food, which seems to be the same as ours. Nobody else has a right to that taste. And people are wasteful creatures. If we prepared only what we need and were careful about cleaning up our crumbs, mice would leave us alone and hunt better pickings elsewhere. The pretty little deer mice prefer the natural thickets to our synthetic ones. They'd rather pat down their tiny trails among the gooseberries and delicately climb the thorny stems for the juicy purple fruit.

Pack rats, the bushy-tailed wood rats, are the big cousins of the deer mice. They are characters. They are fun to know, and it's sad that knowing humans often means their death. They are the size of squirrels and built a good deal like them but are colored like deer mice: brown above, white below, with big eyes, big ears, and bushy white-bottomed tails. They are as avid collectors as people, particularly loving anything shiny. The name pack rat comes from their "packing" objects around, things like pens, rings, brooches, wrist watches. Some call them trade rats because at the place where your watch was you find a pine cone. In pack rat country you hear discussions about his motives. Did he really have such a sense of honor that he went out and got a particularly fine pine cone and brought it back to exchange for the watch? Whereas, if he makes an honest trade, does he feel like a business tycoon? Or is it just that he happened to be carrying the cone when he saw the watch, couldn't carry both and liked the watch better, so dropped the cone for convenience only?

Once Joe, on a pack trip, was sleeping in an abandoned cabin

HOUSES, YARDS, AND GARDENS

near Union Pass. He dreamed of Santa Claus and came full awake when he heard sleigh bells jingling. "How silly," he thought, "no snow and it's a long time till Christmas." But he was still awake when the sleigh bells jingled again. He reached for his flashlight. In its beam he found a pack rat in the middle of the floor, trying hard to stuff one of Joe's spurs down a knothole. The rowel on the spur made the sleigh bell sound.

Around outside the houses, in the somewhat artificial but not synthetic thickets, there are more kinds of mammals. Not many really *live* close to humans. Cottontails do, but we see the others' tracks and sometimes catch a glimpse of them as they go by. Thus we often see the tracks of mule deer and sometimes, in the early dawnlight, a group passing through. Once our downstream neighbor phoned me. "I think I should warn you, Mary, there's a moose in your yard. Be careful." I looked out the window, saw some activity, and answered her, "Thanks for your warning, but he's in your yard now, and the fence is down between us. You be careful, now!" Another neighbor's three-year-old daughter ran into the house one day shouting "oh, Mommie, come see the big dog!" Mommie was shocked to see it was a black bear.

In any accounting of the mammals associated with houses, yards, and gardens, the central and most imposing species must certainly be man. Man and those close associates, cats and dogs. This is not the place for a "study" of man, even if that is the proper study of mankind, but only for a look at his influence on the ecology of my study area.

Before the first white man, other men and women made use of this area. No one knows dates for the petroglyphs on the big rocks in Torrey Canyon six miles south; the local Shoshones say only that they were made by "the old people." My ninety-three-year-old friend, Mrs. Moriarty, says that during the 1890s there was always a winter encampment of Indians on my study area. Dave Williamson heard about the Indian occupancy about the same time. I never knew Dave, but his long-widowed wife Annie, who came here from Scotland in 1906 to marry him, told me the how of it. Dave, with his father and brother Jack, came from Scotland to the U.S. in the late 1800s and worked first on the Mormon Temple in Salt Lake City, then on the headquarters buildings in Ft.

Washakie, on the Shoshone Reservation downriver from us. Interested in the geography of his new country, he asked about the climate up the valley. "Very bad," the Indians assured him. "Storms. Snow. Ice. Nobody lives there. Very bad." But when the winter storms began to be bad at Fort Washakie, the population decreased. Families, with all their packhorses and possessions, were moving out. They were going elsewhere for the winter. And where? It developed that they were headed for the upper valley of the Wind River. A considerable group of them, the ones Mrs. Moriarty had seen, came to my study area. They surely weren't interested in bragging it up so the pushy white man would come there, too.

By the 1890s there had already been too many trappers, prospectors, scientists, soldiers, and homesteaders tramping up and down the upper valley. Colter, Bridger, Sublette, Bonneville, David Jackson, Hoback, Osborne Russell, Lieutenant Raynolds were among the early ones. In the 1830s and 1840s trappers and prospectors were calling it Warm Land and telling each other that its light snow, warm springs, and never-freezing river made it a good place to winter. In 1873 Captain Jones, returning from his "Expedition to Explore the Southern Approaches to Yellowstone Park" crossed my study area. Close to that time the Hayden Expedition camped in it, and William Henry Jackson took photos of the badlands from here.

Jackson's photographs show that this part of the Badlands has eroded very badly in the last 100 years. In particular, the red eminence on the north edge of my area, the goal of most of the Friday walks, has changed a good deal. In Jackson's photo it was a steep grass-covered hill. Only a small area on the southern face was opened enough to erosion so the soft Wind River formation of clays, marls, and conglomerates was visible. Now it is a very spectacular castellated formation, with deep dark fissures outlining fantastic elongated sculptured monsters. There are caves, natural bridges, niches, gargoyles, pinnacles. On this south face there is no pastureland anymore. Even to climb to the top of the structure is a job for stouter nerves than mine. Deer can do it, up a twisted trail of scattered footholds with sheer space between them. With sweating palms and rising neckhair, I can make it about four-fifths of the way up, then I must crawl most cautiously back. How much of this increased erosion is man's influence I couldn't say. I'd guess a large part of it is due to the denuding of the slopes through over-grazing. Probably some of it goes back to the Indian ponies watering here. Most of it has come from white men competing for grass until it is all gone. Some of it may be due to one of the rhythmic pulsations of the Rockies, raising the gradient of the streams. To this last cause at least man can plead innocent: he has no control.

The erosion in a single storm can be spectacular. June 23, 1967, a small cloudburst poured all its water practically at once on this particular hill; hardly any rain hit our acreage half a mile away. But the thick red flood off the south face of the hill was too much for the Locke waste ditch to carry to the river. This water came down in a swift sheet across the land, across our acreage, and down around our river-house. A natural levee kept it from going into the river, and it stood a foot deep in places. It reached through the gravels and colored our water supply pink. It took two days to dig a trench through the natural levee to carry the drainage to the river. The red flood drained away, leaving a gummy half-inch of mud in the yard. The next walk to the Badlands showed me visible changes in this one hill. The channel under a natural bridge had disappeared and new ones had appeared. The figure sculptures and gargoyles wore new expressions. The deer trail was even scarier than it used to be. The gullies draining the lower slopes were little sheer-edged canyons now, in places uncrossable. Angling across the approach lane were new lens-shaped sorted deposits of sand, gravel, and red mud patterned with ripple marks.

I can't prove it, but I'm convinced that if man's horses and cows had never overgrazed the hill, much of this storm would have soaked in or found its way harmlessly and more slowly downhill between the plants. And how many such storms must have violently attacked this bared-off hill in the last hundred years?

The effect of his favorite animals, dogs and cats, upon the various habitats is less extreme than man's own. Loose dogs will occasionally band themselves into packs and chase deer, cattle, or horses. If they chase man's animals, the owner of the stock will shoot the dogs without compunction. If the dogs chase deer, game wardens have to be more circumspect than cattle ranchers—the shooting can't be openly done. In my study area I don't know of its being done at all.

Cats are different, especially for an observor of birds. It seems that cats can never be trained not to catch birds. But to wildlife, even birds, cats are not the universal enemy that man is. This one species has contrived to make himself feared and hated by most other creatures. Since this fact is rubbed into my consciousness day after day by many creatures with whom I would be friends, I grow sensitive and ashamed of being one of such feared and hated beings.

But there are a few who, knowing man, hate him not; a few who accept man as one of themselves, part of the community or body of all life, who manifest toward man tolerance and understanding. Silly creatures, perhaps, but very winning from people's point of view.

Kestrels, the assured little falcons of the meadows, consider themselves quite capable of carrying on a friendship with man. I've known several kestrels who lived in apparent contentment with friends of mine and hunted for them as trained falcons. Mice, sparrows, and grasshoppers were their prey. Since man is a predator, like a falcon only not so stylish, there's a community of interest here more appealing than that between man and deer mouse, where the likeness is in taste for food.

Mourning doves will come to the houses where bird seed is thrown out and will walk about like moving gray stones as they pick up their meal. They are ready to go on split-second notice, and you are suddenly aware of the V-shaped white accent pattern on the edges of the gray tail feathers.

Hummingbirds, tiniest of birdkind, are famous for coming to gardens, but I have seen few of them in my study area. One must have feeders supplied with a formula of colored sugar syrup, I am told, and so far I have failed to set one up. The one I see most often is the rufous hummingbird, bright reddish brown with brilliant red throat. I've also identified the black-chinned hummer with black and purple throat, and I see with them the confusing green and buff females.

If you live in the woods you have the woodpeckers for neighbors:

red-shafted flicker, red-naped sapsucker, hairy, downy. They pay little attention to people, whether to like to dislike them, but they are very visible. They drum loudly and scream loudly.

The western wood pewee, ghost of the cottonwoods, is a famous mosquito chaser who is good to have around. His presence does not signify approval of man or his works—only pleasure that man's naked skin attractes so many mosquitos. He will hang around the river-room when people gather and a picnic is in progress. On the table will be a tube of repellent, and the picnickers will be applying it generously. Sure enough the insects are first attracted, then repelled, which makes for a sort of confused mob action most satisfactory for the pewee. The little gray bird sits upright on a twig and makes so many sorties and returns, always successful, that he hardly has time to let out the lonesome "Prweeee-p" that tells the world he's on the job.

Swallows, like pewees, value people for their insect-attracting powers, and from ancient days people have appreciated swallows for their enormous capacity to eat insects. And swallows have always been admired for their flashing flight, their brilliant color, their expert masonry, and their pleasant domestic life. The three species living close to houses in my study area are violet-green, tree, and barn swallows. Tree swallows live in downy holes near our river-house; barn swallows nest in the porches of the Red Rock Ranch Motel. I don't know where the violet-greens live, I know only that the young appear in early June on the twigs of nursery-trees by the river, the green and violet looking crudely daubed on, the white all fuzzy, the general appearance pert but unsure and questioning.

You might say that all the birds who make a policy of taking advantage of man's presence are opportunists by nature. The greatest opportunists of them all are magpies. Like man, they are pushers and shovers, investigators. They are impudent, cocky, sure of themselves, and very social. All these qualities are bound to bring them out of the thickets and around man's domiciles. They are often raucous and noisy. The trouble with them is, they mock man and his works, often to the point of enraging an individual man. So he resolves to shoot them. It's not hard to shoot the first magpie but almost impossible to shoot any more. They get the message the first time and then make a point of avoiding direct encounters, but with alertly sophisticated mockery. I think that in the interest of sanity it's better to get along with magpies than to fight them. I chuckle at the assurance of a perfectly groomed mapgie, black head high, eyes gleaming, white scapulars and underparts shining, green and blue iridescence on his long tail feathers, as he struts across the yard near the feeder. Sure, he does eat some suet and some sunflower seeds I'd prefer to have other birds use. Sure, he chases other birds sometimes. Sure he eats robin eggs when he can get them. Sure, he jeers noisily at anyone who puts on airs of superiority—eagles, goshawks, or man. But there's grace in admitting that man is fallible in these ways too. Magpies can enrich our experience of brotherhood.

The black-capped chickadee, my favorite bird of the thickets, is a favorite at the feeding station, too. Always cheerful in the grimmest weather, hard working, alert, quick, he is a sure lightener of the spirit. Anybody with a feeding station will have chickadees if no other guests, yet they're not primarily moochers. They like to get off welfare and make their own way. They eat from your bounty because they're invited: they know you want them to. They enjoy your charity, but they aren't bound by it; they still work the thickets for insect life in all seasons. At the feeders, chickadees like peanuts, sunflower seeds, peanut butter, doughnuts, suet. They aren't greedy. So much of their time at the feeder is spent in looking around that you wonder sometimes if they're eating enough to get along; but the eating itself is efficiently done. A bird carries a sunflower seed in his beak to a twig away from the feeder. He puts the seed between his slender black toes and grips it against the branch. With sharp little pecks and twists of his beak, alternating with raising his head and looking around, the husk is worried off. Then the meat of a sunflower seed is four bites for a chickadee.

The mountain chickadees with black-and-white striped heads sometimes come to our house in winter. The Shoshone name for chickadee, "Widg-i-gee," I think originated with this species, whose call is lower pitched and huskier than that of the black-cap.

If the black-cap is the favorite bird of the thickets, there is no question that white-breasted nuthatch is the favorite bird of yard and garden. A pair of these droll little trolls have kept us happy for the last eight years.

Before we built the river-house, we knew the sound of the nuthatches' tiny tooting at the site. Sometimes we would see them trotting around upside down on the trunks of the cottonwoods and hear the dry lisp of their claws on the bark. At the time we started building we established the suet-tree, which would be just outside the kitchen window when there was a kitchen window. The tree was a mature cottonwood about eighteen inches in diameter. We fastened a chunk of beef suet to it with a nail and a length of baling wire. There was just about time to back away when the first visitor appeared. A scratchy sound of claws on bark, a little "Toot-toot!" high up, and a small bird spiralled down the tree head-first, cocked a beady black eye at us, pried off a penny-sized chunk of suet, gripped it firmly in a sharp black turned-up bill, and scooted out of sight on quick wings.

He's a little fellow, less than six inches from beak tip to squared-off tail, and very plain of color, but you could never call him unassuming. He truly assumes. Both male and female are gray above and white below, with an all white face. Black accents are bill and feet, markings on wings and tail, and bright eyes. The male has a shiny black cap, the female a charcoal gray one rimmed with black. It's a sober outfit for a bright personality.

All the year-round for eight years two have come to our feeders almost daily, so we're bound to think they're the same birds. Peterson's says that they "wander in the winter." Well, that accounts for the occasional winter stranger that gets in a fight with our stay-at-homes.

They are addicted to people-watching. We are their zoo and their TV programs. Joe says that they are writing a book on people. After eating their fill, they will often fly from the feeding shelf to the suet-tree, sit on the bark upside down and turn their heads to look directly at us. It's a curious thing to see. The beak is a small black diamond when seen head-on, the sharp black eyes are uncannily close together and close to the diamond-shaped beak, and the big white face flares from that tiny center to the full width of the shoulders.

The odd name of nuthatch was given these birds, not because

they brood nestfuls of nuts or hatch young from them, but because they use their beaks as hatchets to split nuts. The first time I brought whole peanuts the nuthatches were puzzled. After rolling one over a few times, the male backed it into a corner, held it with a foot, and drilled a hole in one end, using sidewise thrusts of his beak. When the hole was big enough, he stuck his slightly upturned beak into it, and flew with it to the suet tree, where he jammed it hard into a crevice in the bark. There, at leisure, he demolished the shell, pulled out the peanuts and ate them. We have as one of our woodpile tools an implement called a picaroon. It's a single-pointed steel tool about ten inches long with an oval eye in the other end into which an axe handle fits. You whang this down into a log, and it makes a handle to move the log with. With the aid of a picaroon, even I can shift a twenty-foot log with comparative ease. When Joe watched the little male nuthatch drive his beak hard into a nutshell, he named the little fellow Picaroon. Naturally his wife then became Picarette to us.

Picaroon and Picarette are inseparable, but he is definitely the boss. Or perhaps, as you may see in human society, she has found she can improve family morale if she kowtows to him as boss, and it doesn't hurt her any. Often you see them upside down high in the treetops, hitching their way down parallel branches, and hear them visiting companionably with short whistled phrases, putting a lot of expression into "Whi-whi-whi!" Sometimes he will find a good grub, summon her with a sharp "Toot!" and present it to her with a short nod of his head.

At our feeder he is quite the show-off and domineering male. They seldom come exactly together. If he is on the feeding shelf she is on the suet tree at a slightly lower level or watching from farther off on the porch roof. If he is above the suet, she is below. If he shakes his head at her, she ducks quickly out of sight arund the tree. I felt sorry for her for a while, thought women's lib should do something for her. But she set me straight. She can take care of herself without help. Just one time he shook his head at her once too often on the suet tree. She sat back on her tail and glared at him. Every feather on her head and back stood straight up, making her a tiny caricature of ferocity. She cupped her wings, quivered all over, and dove at him. He was quite literally taken

aback. He jumped back two ridges of cottonwood bark, pulled back his head, and looked down his beak at her. He made up his mind, went to the suet, pried off a sightly chunk, hitched over to her, and presented it to her with a little bow. She took it calmly as her due, smoothed down her feathers, and the war was over. He didn't shake his head at her again for quite awhile.

Every year they raise a family. During the time she is brooding eggs, she never comes to the feeder, and Picaroon keeps her supplied. Until the young are a-wing, both the parents work frantically to quiet their enormous appetites. Finally each year the proud parents bring their brood to a coming-out party at the feeder. More than once our first view of them has been the sight of a row of six heads and six sharp beaks, upside down, looking at us from the edge of the eaves. Peterson says a clutch of nuthatch eggs number six to nine, but three or four grown youngsters seems to be the rule with Picaroon and Picarette.

The little mites of house wrens are as noisy with songs and scoldings close to the house as they are in the thickets. For three years a pair has raised a family in a hole above our outdoor lunch table by the river. The parents sing and scold and bring bites to the babies while people eat just below them. We give the wrens full credit for courage, toughness, and hard work at reducing the insect population, but we don't find them nearly as winning as nuthatches. In part, we must blame it on the fact that their season with us is our busy season, while we get to watch nuthatches when there's more time for observation. The house wren's pleasant warble first sounds in late April or early May, and my latest record is September 27.

Way up at the top in almost anyone's list of favorite garden birds is the robin. This cheerful thrush has it all—good looks, bright color, sweet song, confidence, poise, friendliness, domesticity. It's curious and flattering that he should enjoy people's company, for he's no moocher. What humans set out on bird feeders is not for robin; he finds his own. He can get along well without people. I was astonished to find robins near bivouac back-camps high up in the spruce forests, and they were flying across the tundra when we camped on the banks of the Yukon River. Without being nosy, he seems to hunt out humans wherever he is, and proffers his friendship.

Robins love water. They drink a lot and bathe a lot. They sometimes follow the hose sprinkler around as I move it over the lawn, once four of them together. In a chilly April rain, I saw one fluffing out his feathers on the lawn, deliberately taking his cold bath. On a hot July 2, a robin in the yard cocked his eye at me, flew to a log end close to the window where I stood, opened his beak and panted, looked down at the dry bird bath, then at me, and panted some more. I went out and filled the bird bath. He hardly let me get it done before he was in it.

One July 10 I watched a curious happening. A female robin had a blue robin's egg in her bill. She flew to a branch about ten feet above the ground. She paused, peered around, then *threw* the egg with emphasis toward the ground. Was it an infertile egg? The egg of a rival? The egg of an enemy?

August and September are the times for consolidating learning, doing things as families, eating lots of gooseberries and spreading lots of purple manure around. By late September small family groups are watching the weather. Storms from the west in October will launch most of these little groups on an eastbound trek away from here.

The other dooryard thrush is mountain bluebird, the loving little sweetheart. The all-blue mountain bluebird and his blue-trimmed gray mate are desert birds. The species belongs in the next chapter, but I have been privileged to know two bluebirds and their offspring as dooryard friends and so must tell about them here. When we moved from the studio on the highway into the shadows of the cottonwoods, we gained many things; but we lost the bluebirds. Their whole system is delicately adjusted for life in the open. Thickets and woods are scary to them. As long as we live in the woods, we'll no longer wake in the summer mornings to the soft domestic music of bluebirds telling each other of their love.

It was early May in 1959 that a pair of bluebirds who had been hanging around for a month or so began seriously looking for a place to nest. Together and separately they explored both gables and the log ends of all four corners of the building. They hunted for other nooks and crannies: where the yard lights are attached, and all along under the eaves. If there were only a hole big enough, I'm sure they would have been in the attic. A birdhouse was indicated. Joe and I were more than busy getting ready for tourists; besides, I was teaching art every weekday in the Dubois schools. But the bluebird calls began to sound frustrated and complaining. I found a length of 1x6 board around six feet long, did some quick measuring, cut it up with a hand saw, and in a short while had a birdhouse of sorts nailed together. Where could we put it where we could see it? The only choice in our desert setup was the power pole in our back yard. Would bluebirds accept it and its hum? The answer came fast. After just one excited, fluttery look, Mrs. Bluebird staked her claim with a piece of grass.

From that time on we had bluebirds all summer every summer. Often several arrived together (the old couple, plus some of last year's children?), but only one pair stayed, though we provided a second birdhouse.

The usual routine was: arrival in late March; courting, playing around, and surviving snowstorms all April; nesting, beginning early in May, involving both mates bringing in nesting material and the fussy housewife's discarding most of her husband's offerings; hatching in June; about seventeen days of frantic insect-carrying between hatching and flying; then a whole summer of education for the youngsters. Usually part of their education was watching the progress of a second family of brothers and sisters. By early July the female would be brooding a new clutch of eggs, the male being very attentive, looking in often and sweet-talking through the little birdhouse door. The flying youngsters mostly learned their own lessons and adventured more widely. By mid-July the adults would again be feeding young; incredibly to me, the first family would come back and help—not seriously but gaily and irresponsibly like any teenagers. The new family grew up, took wing by August first. From then on, a close family group took part on all activities, bathing, grooming, going on walks with me, exploring the bridge, visiting on the wires, hunting insects. Then the children left first, some time in September. Mother and Father had a last honeymoon, a final fling, and left sometime in October.

Lots of times the routine varied from the one described, variations due to changes in the weather, the health of the bluebirds, intrusions by other species, and individual reactions to the educational process. I could write a whole book on bluebirds, supported by a hundred index cards written small and packed full. I'll select just a few incidents that stand out in memory. On April 7 with snow falling fast, a male bluebird sat on the elk horns nailed above the kitchen window, a chilly little fluffed-out blue ball. His sharp black eyes looked down at a group of busy rosy finches eating seeds laid out on the top of Joe's tool box. I sure wished I had some ants or mealworms for him. He flew down to the tool box. The finches gave way, and he tried twice to eat a seed, spat it out, shook his head violently, and flew away. Joe asked me, "Didn't you hear him say, 'Hell, *nobody* can eat that damn stuff!'?"

They lived through that cold snap. By May 7 they were nesting in the house on the power pole. On the 26th both adults were carrying in insects and worms. On May 28 Joe saw them mating again. They were apparently feeding young up to June 14. We didn't see them near the birdhouse on that day or the next two. On the 16th I saw both adults, not near the birdhouse but on the other side of the highway; the male was delicately, with fluttering wings, feeding the female. But what about the babies? By the 17th we were fearing a family tragedy.

So Joe got the stepladder and a flashlight and took a look. Amazing! What he saw was a dead chipmunk just inside the entrance hole. Joe and I finally pulled him out, using long-nosed pliers and many false tries. He was a tight fit in the door. We theorized that the chipmunk had invaded the house while the female was on the nest. She lashed out fiercely with her bill in defense of her family. Chipmunk turned to flee, but there was Father even fiercer in the doorway, and between them they pecked him to death. Talking over the strange situation, we decided the parents had

been too upset to stay, had no way to remove the chipmunk, and had abandoned the babies.

We had to check. I climbed the stepladder, Joe brought me a hammer, and I pried off one side of the roof. If I'd been dismayed before, that was nothing to the shock this time. Far from abandoning her nestlings, there was Mother Bluebird sitting tight, very much alive, bright eyes fixed on mine, not moving except for a fine trembling that vibrated the little house. Imagine! A clumsy giant going "Screech! Scraunch!" with a big implement, actually taking off your roof, then peering in at you with eyes as big as your head! Terribly embarrassed, I hastily tried to put the house back together without banging with the hammer. I tried to push the nails back in the same holes. Best I could do, there was a big crack down the middle of the roof, and it was starting to rain. Joe cut a scrap of tar paper, and we thumbtacked it on to the roof. And do you know, the bluebirds raised that family!

Sometime later there was a follow-up to this story—a mixture of sharp chipmunk alarm call with sharp bluebird talk. Joe and I came running, for bluebirds can't talk sharp. There was a chipmunk tearing across the back yard and away from the power pole, with a male bluebird hovering close above his back and delivering good hard pecks.

One April day an amorous flicker discovered that the birdhouse made a beautifully resounding drum for his mating call. He drummed loud and long, trying the top of the roof, the front, the side, then the floor from below. Meantime the rightful owners sat all ruffled up on the pole fence nearby, and at frequent intervals flew and dived at their enormous tormentor. He would cock his head at their harmless sallies and just deep on drumming. It was all of half an hour before he turned the little house back to the bluebirds.

The most dangerous days in the life of young birds are the first few after they leave the next. No wonder they hesitate. Their flight equipment is rudimentary, not yet aerodynamically sound, and they've had no training in its use. Instinct is all right, and it's all they've got at this stage, but nothing beats training and practice. I saw one unlucky little mite on his first flight from the next land sprawled in the side road and immediately get squashed

by a car. In another first flight, by contrast, the mother led three little ones from fence to fence across that same road and back again. It's easier for untrained feet, wings, and eyes to make a landing on wide barky wood than on wire. At night the children want to go back into the house, but Father and Mother want to honeymoon in there, too. I saw three little ones as dark came on, repulsed from the house, sitting glumly side by side on a step of a ladder that leaned against the fence, soft baby mouths drooping, bits of white down sprouting crazily between the gray feathers. Father Bluebird took pity, flew past slowly, talking to them in chuckly bluebird language. They followed him, flying across the yard and into a leafy willow by the ditch, where he established them for the night.

The event that nearly broke our hearts was the bluebird disappearance in the middle of nesting season. We might have been warned. That year Mother seemed a little lax in her duties: one evening in the middle of May, Father stuck his head in the nest box and warbled briefly. We thought he was making love-talk with his wife, but no, he was just discovering she wasn't there. He went to the edge of his porch and warbled with urgency: up came his lady-love from the ground and slipped inside. He had caught her neglecting her chores. Two weeks later the weather turned cold and rainy. At 2:00 P.M. we saw the male fly to the nest box. He looked in, looked around from the porch, and called and called. Finally his mate arrived, all flustered and protesting. The next day I saw Father take an insect into the house, then take it out again. It was the last sight of a bluebird we had for quite some time.

What happened? Why had they left? We were afraid some predator had got them. We desperately missed their sweet voices. But tourist season had started with more than usual pressure; we had no time to investigate; we did our jobs in the studio and just silently mourned.

Two weeks later, at mid-day in mid-June, we heard a bluebird warble. We both left what we were doing and ran to the kitchen window. There were a pair of bluebirds at the nest box. One went in, came out. The other went in, came out. They flew nervously to the fence, to the phone wire, to the light wire, back to the birdhouse. The male came to the pole fence at its closest point to the house and looked in the window we were looking out of. He stretched toward us and talked urgently. (We thought he said, "Do something! Can't you DO something?") Joe set up the stepladder, pried off one side of the roof, and lifted out the nest. He handed it silently down to me. There inside it were four dead hatchlings and three eggs besides, one of them pipped, all very cold. I took away the sad debris while Joe replaced the roof. Both adults were watching from the phone wire. So that told the story: flighty Mama stayed away too long on a cold day. She tried to brood life back into dead babies, and it couldn't be done. Father tried to get the dead babies to eat, but it couldn't be done. So the parents sadly gave up, moved out, decided to try for a new life somewhere else. But something brought them back to us. What would be next?

The next day, Father took a small moth into the bird house, then took it out again. Pitiful. I gathered a bit of grass and a few fibers of sagebrush bark, and pushed the wad into the entrance hole. Both birds accepted it and worked together to push it on in.

We were in the bluebird business again. This time they raised a family, all on the wing by July 26. This time Mother was a model of propriety. Nobody ever explained the minor irregularity of two babies being on the wing a full week before the last two left the nest.

Anybody with a feeder and berry-bearing shrubs will attract a considerable list of visitors. Some of them I've described in other habitats: Bohemian and cedar waxwings, shrikes, starlings, yellow and Audubon warblers.

Now comes the house sparrow. Over most of the country he's the one bird most people know by sight. Most would say you find him everywhere. This is not really true. House sparrows have become so dependent on people that they seldom live away from them; so when you are away from concentrations of people you seldom see them. I came to the Wind River Valley in 1935, and until 1946 (except for a war-work interval) Joe and I lived on mountain ranches high above the village of Dubois. In those years I never saw a house sparrow on the ranches and very seldom when I went to town. In 1946 we moved to our present acreage two miles below Dubois, and in 1947 the first pair come to live with us. Since then they have always been around, winter and summer, but in no great numbers. In winter there may be a couple of dozen, and they mostly disappear in spring. Perhaps they succumb to disease, exposure, or predators, or simply look up new families to live by (for in our valley as elsewhere there is a population explosion). For at least six years we've begun the summer with three or four pairs of house sparrows in our thickets. Each pair produces several families before winter. There is a rather high incidence of infant mortality, but we get our two or three dozen for winter again.

I was used to house sparrows in Chicago and its suburbs, where I thought of them as slightly troublesome, dirty birds, dull-colored and uninteresting. Apparently city smog darkens the light feathers and subdues the contrast, out here in clear air I've discovered the male's bright pattern.

Because of the sparrow's close association with mankind, most of my contacts with them have been at my feeders or around the woodpile. They are essentially thicket birds. They particularly like the thick mass of willows, birches, and silverberries just upstream from the river-room. Here they gather in noisy conventions on winter mornings at sunrise. They seem to enjoy doing their toilets in company, conversing shrilly as they shake their wing and tail feathers into place, and run their bills down secondaries, primaries, and coverts. The sun gives warmth, and there is shelter from the west wind. They hate to move when I come along, and they won't move for the dogs. They aren't as numerous or as noisy in the thickets in summer, but the ones around are still in the same patch of brush at sunrise.

Grackles, cowbirds, black-headed grosbeaks, indigo and lazuli buntings, evening grosbeaks in winter, all come to the feeders and sometimes hang around the yards. Sparrows who accept welfare handouts on occasion include pine siskin, green-tailed towhee, slate-colored and oregon junco, tree and chipping sparrow, Harris' and white-crowned sparrow, Lincoln's sparrow, and our delightful friend of the thickets, song sparrow.

The rosy finches deserve a chapter to themselves. They are interesting for their pioneer spirit, their foolhardy love for the strenuous and most difficult country. When they show up, you know it's completely impossible to live on the mountain tops. Our high valley is what they call "going south for the winter," and they hate to admit that they're that chicken. Because of this quality of theirs, most people never see rosy finches at all, and for most of those who do see them the occasion is so rare they don't know what bird they're looking at. Yet the little birds aren't shy, and when you see one, you generally see many.

They nest only in bare rock above timberline. In our latitude, that means above 10,000 feet, though in the Brooks Range of Alaska I understand it means a wide spread of altitude down nearly to sea level. They stay by choice close to perpetual snow.

All the rosies dip in the air to the beat of their wings, somewhat as the horned larks do. Both larks and rosies resemble wind-drifted leaves as they fly—the light birds are larks, the dark ones rosies. Completely birds of the open, still they come eagerly and by the hundreds to my feeders in the woods to eat seeds and suet. They are most likely when winter storms are bad. When each storm is over, as soon as the sun comes out, up the long mountain slopes go the little dipping flocks.

The Desert

ALL THE LIFE ZONES, habitats, or ecological communities—whatever you call them—that I've been looking at so far are blessed with water. But half my study area is desert. The average annual precipitation in Dubois is only about seven inches. In selecting my study area I plainly avoided typical western Wyoming when I picked a river valley well-watered beyond reason.

We are in the "rain shadow" of the Rockies. The ridge of the Continental Divide curves around the head of our valley, fifteen to thirty miles from us to the south, west, and north, at an altitude of 10,000 to 12,000 feet. Sixty miles west of us are the Tetons, a 13,000-foot-high comb against the clouds, standing north and south. These two ridges catch the clouds in their unsteady movement from northwest to southeast, and they milk them all, squeezing out their moisture so that very little falls on us. Our salvation in this water demanding life is that Wind River and its tributaries begin high up in the cloud zone. But those streams don't affect *my* north and south deserts.

The north desert is the edge of the Wind River Badlands, fantastically eroded red-and-cream-striped hills. There is a valley called Mason Draw, a dry wash that carries water only during rainstorms. It drains Table Mountain six miles north. It probably has some permanent water deep underground, because here and there in its bed is a sand-bar willow or a scraggly cottonwood.

The south desert is the beginning of the foothills of the Wind River Range. The east edge of my study area is the tumbled moraine left by Jakey's Fork Glacier, the west edge is the frayed-out end of Sheep Ridge, and between is the alluvial fan of CM Draw, with the narrow fan handle of CM Ravine cutting the hills at the extreme south. CM Draw is as dry as Mason Draw. One or two stunted willows and a bush or two that might with luck and water some day be a cottonwood—only these whisper of any underground moisture at all. It looks as if the story told by the visible erosion has got to be a lie.

But every few years comes one big storm, and erosion under your very eyes attests to truth. The heavens are blackened, lightning snaps and thunder crashes all over the sky, water pours down in such masses that all one can do is cower in a safe spot. After hours, or rarely after a whole day or two, the clouds show bright blue rifts, and I dare venture out in slicker, boots, and plastic rain hat, to check the sodden landscape. In Mason and CM Draws, there are new patterns. In Mason Draw the channel has changed, there are new-sorted lenses of fine and coarse sand and gravel; banks have been eaten into and caved in; there are piles of sagebrush debris here and there; channels of sticky red clay come steeply down from new sculptures on the badlands walls. In CM Draw the sharp ravine that is the handle of the alluvial fan has deepened. Cabin-sized granite boulders have been shouldered many feet downstream. Tributary draws are more sharply cut.

Downstream from the ravine's edge is the great change. Just out of the ravine at the pole gate, the water has abruptly lost its carrying power. Instead of cutting, it is now filling: that's the essence of an alluvial fan. The big storm is making a vivid diagram of it. I had seen before that the fan was higher in the middle, all the way from the mouth of the ravine half a mile or so to Wind River, including the Red Rock hay meadow and pasture. Where I'd expect a stream channel and the lowest part of the valley, there I'd see a gently upsloping ridge. Until the storm it didn't explain itself, but now the process is clear. During the pressure of the big storm, with water everywhere, the big push coming out of the canyon drops its loads of stones and boulders at the pointed top of the fan. But more water splashing from the sky brings more power. Stones and smaller boulders go farther down the draw, on top of the central ridge, and more and more stuff keeps coming up from the ravine. Gravel drops out further down the fan, sand is carried almost to the line of the highway, and silt is spread out at the bottom. The skies will clear. The new sand and gravel bank will keep the middle of the fan high. New growth will cover it next year. It will be years most likely before another such storm will rearrange the fan.

The narrow basic valley which is the scene of the action of the alluvial fan dates back to the Jakey's Fork Glacier. Before it and before the Ice Age that mothered it, a drainage channel or ancestral valley took water straight east into Jakey's Fork. Its forking upper channels are still intact. But then came the long cold years when the valley of Jakey's Fork gradually filled with ice, and the ice flowed shuddering north toward Wind River, scraping up great wads of dirt and rocks on its underside, gradually working them sidewise and out on its flank. The dirt, rocks, and ice dammed the flow of the ancient side stream; the water found its way down a new valley between the rough moraine and the stream-laid beds of the old Wind River Valley. This must have been going on for hundreds or thousands of years while the moraine built up higher and higher, then while the ice slowly melted back and the glacier disappeared. In CM Ravine you can stand in the narrow sandy channel, spread your arms wide, and almost touch the very different walls. West, the once-in-a-while erosion is cutting into level layers of stream-laid clays, sands, and rounded gravels. East, the wall is a ragged accidental pile-up of rock debris: polished boulders, sharp-edged broken pieces, gravels that shift from pebbles to clay in an area you can cover with your hand. Many of the boulders are big as houses, and they stick out of their rough matrix like raisins out of a bread dough waiting to be kneaded.

The variety of rocks and minerals in Wyoming has encouraged the development of a host of rock-hounds. My study area is not a particularly rich field for them, but it's not without interest. The most barren spots are the most attractive, because the stones can better be seen without competition from vegetation. These spots are the freshest gravel bars in the river, the highest part of the alluvial fan, and the driest parts of the desert.

Most eagerly sought are agates and agatized wood. In past ages there was a lot of silica in the very hard water, and the surplus was deposited whenever the water slowed. When the mineral-loaded ground water finds a hole underground which it can fill quietly, the excess silica precipitates all around the walls. In time it may completely fill the hole. This mineral silica is found in my study area in four different forms. If the hole which is filling with deposit has a steady and continuous source of water, the deposit may be a milky white, very hard material without grain or pattern, called chalcedony. If the source of water is not steady, but dries up at intervals, and carries material other than silica, the deposit will have seasonal rings almost like the rings of trees. Impurities will add color and pattern to the piece; sometimes the chemicals present will arrange themselves to make a beautiful design, or a landscape with trees, mountains, and water. This is agate. The third form of silica is quartz crystals. These may line the center of the hole being filled, several layers deep. The fourth form of silica is sand, which is any of these three forms ground up and battered to pieces in the powerful streams of spring.

The holes in the earth which become filled with one or another of the forms of silica are quite often made by the rotting away of buried twigs, branches, or whole logs. Sometimes the wood debris is all carried away before depositing starts. In this case the deposited material will have no resemblance to wood structure. Only the surface will look like wood, with knots, grain, and bark. Such pieces the rock-hounds call "limb casts." Sometimes, though, the water carrying away a single molecule of wood will be followed immediately by water carrying a single molecule of silica, which it will stuff into the molecule-sized hole. If this continues, molecule by molecule, clear through the piece of buried wood, the agate will be a complete and accurate study of wood structure, with tree rings and all details of grain, inside as well as on the surface. This is "replacement wood" and is highly valued.

In my study area the depositing water is no longer rich in silica, so agates are no longer made. Instead, calcium and magnesium are the common minerals in the ground water. Buried wood is right now being turned to stone, but not to agate. The method is the same but the product is not so pretty. It is called calcified wood. It is a medium grained soft limestone or dolomite depending on whether it contains more calcium or magnesium. The deposition takes place so fast that once in a while you may find a stick that is wood on one end and "stone" on the other, a very soft unconsolidated stone but all mineral and showing the details of bark and knots on its stony surface.

THE DESERT

In Mason and CM Draw, all these forms of petrified wood may be found where the rare storms work over what was already there and bring down more specimens from higher up. All summer occasional rock hounds walk the dry channels, every now and then finding a beautiful piece that may be sliced thin, or carved into lovely shapes, then polished, showing deep rich colors, swooping grain, and astonishing patterns.

The best material is Oligocene in age and dates from the long volcanic period when the Absaroka Mountains were being piled up. I have read that a geologist identified eighteen successive forests in the layers of one Absaroka cliff. It makes the imagination reel to try to understand the length of time involved. For instance, I've tried setting out young spruce and juniper trees and found I'm lucky if I can see two inches of growth of year. As another example, the Jakey's Fork Fire, a 1200-acre scar on the slopes of Whiskey Mountain overlooking my home, has been a blot on the landscape since 1934, almost fifty years ago; it is barely beginning to grow new trees; our children's children will still recognize it as a fire scar. But at what more primitive stage would it be if, instead of a forest fire, it had been consumed and buried under hot lava and volcanic ash? The fresh-looking lava flow in the Craters of the Moon National Monument is supposed to be several hundred years old. To grow a forest, then bury it under volcanic rubble, then grow another, bury it, and repeat eighteen times—it's incredible!

Finally, imagine these eighteen successive forests all turned to agate; and imagine all these layers of agate being straight above where I stand, occupying space in what is now the clear blue sky of the Wind River Valley (only a little polluted by the burner of Louisiana-Pacific). This is a real struggle of imagination. It is from those thousands of feet of volcanic debris now washed away that the agates found in my study are derived. They are much heavier than the rest of the sediments in which they once lay, so they managed to hang back when the rest of the mountains were washed downstream. In the long course of the relentless years, though, even most of these heavy pieces have washed downstream out of my study area.

Sagebrush and grass are the conspicuous forms of vegetation in the desert. The Forest Service says that the sagebrush used to be much less prominent than the grass, that it was mostly confined to the draws, and that generally the ridges and the flats were covered with rich grass that might grow knee-high. The rangers say that when the desert grass is overgrazed the sagebrush extends its range and that the overall spotted pattern of dark sagebrush with light grass between is a recent phenomenon. My experience doesn't go back that far, and neither does that of the rangers, so whether this is truly so I don't know.

My north and south deserts are not grazed all year round. Right now, no cattleman is using the south desert. The north desert is part of the pasture of the Diamond D Ranch, whose headquarters are miles northwest. Their cattle or horses seldom roam this far.

The grasses of the two desert areas are collectively called bunchgrass: they grow typically in small bunches separated from others by bare ground. Depending on the available moisture the circle of bare ground may be an inch or so to a foot or so wide. Under the ground the eager roots explore all the space vigorously, extracting every particle of food and water. Not often do you find anything like a sod with interlacing roots. In good years and with limited grazing (sometimes in the Mason Draw area) the bunchgrass will look like a hayfield, with the tops interlacing, higher than your boot tops in the best places. Usually, though, you can see the halo of bare, tan ground all the way around each bunch. Precisely because each bunch contains all the foodstuffs from a large area, the slender blades and firm little grains in the heads are especially nutritious. Cows fattened on western Wyoming bunchgrass often top the market in Omaha.

The best cattle food is the group called wheat-grass. I've found five species of wheat-grass in the two desert areas. The richest is desert wheat-grass, with heavy curving flat heads shaped like arrowheads and full of firm grains. It grows in the best of the desert soils. Blue-bunch wheat-grass is another good one, with long rough-bearded heads. Western wheat-grass grows on alkali land; it has thin stems and narrow close-set grains. Bearded wheat-grass is much like it with grains a little larger, each with a short whisker. Slender wheat-grass, found on light sandy soils, is the skimpiest one; there is a space of thin clean stem between each two grains.

THE DESERT

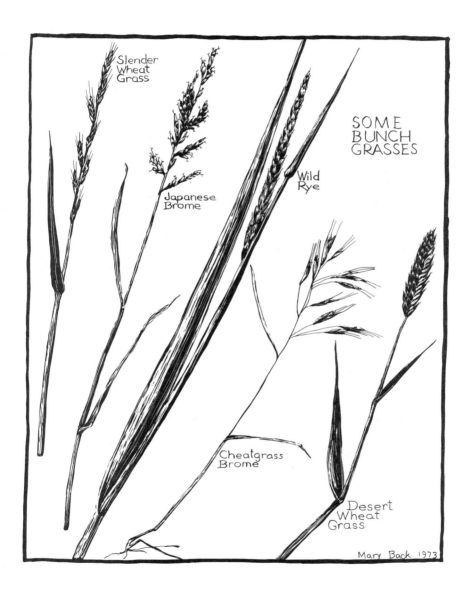

Indian rice grass is one of the commonest desert grasses. The head is a loose open network, each little round grain widely separated from its fellows on a fine stem of its own angled away from the others. The plant grows about eight or ten inches high; the lacework down there around your feet, green in the summer and tan in the fall, is a kind of magic.

Canby blue grass makes clearly recognizable blue-green islands with narrow blades and loosely firm heads.

Needle-and-thread is as odd as its name. The grains are long and narrow (needles) and widely separated on a nodding stem. Each grain has a stemlet of its own, and it also sprouts a long thin hair (thread) perhaps as much as five inches long, all the threads twisting and curling around each other.

There are other grasses. The beautiful and deadly foxtail gets along in places in the desert hills. Then there are the big tough banners of wild rye in places on the alluvial fan of CM Draw or in some of the dry ravines. It's our biggest grass, sometimes six or eight feet tall, growing in bunches a foot or two across. Rye is a very decorative grass, with long strong curving blades half an inch wide and heavy rich seed heads. The leaves are too tough and coarse for horses and cattle to eat—it's considered a sign of starvation when stock is eating wild rye. But chipmunks and picketpins like the fruit; they cut down the stiff stalks like trees in order to get at the seeds at the tips.

Grass and sage—that's the impression you get of the dry country. But as you walk it in all seasons you become aware of great numbers of other plants living there, too, each with its own beauty, its own problems, its own solutions.

Right on the rocks, for instance. Those big house-sized boulders balanced on the moraine slopes or resting between storms on the alluvial fan—they aren't bare rocks at all. They are almost completely covered with a crust of lichens, coloring the granite or quartzite or diorite with the brilliant colors of their bodies. Far off in time, and up to this very day, lichens are the first agents, after frost, to work at turning rocks into soil. They grip the mineral surface tightly with their whole bodies and secrete an acid that digests rock to get part of their food. The algae part of the creature, besides, has the power most plants have to manufacture sugar out of carbon dioxide and water in the presence of sunlight. They store this sugar, combine it with other plant food from the rocks, and build up their bodies with it. When they die, those bodies contribute a tiny bit of dust. This dust is enough soil to let more complex lichens start up, lichens that look a little like gray-green leaves. Where frost has formed a crack, and into the crack has fallen dust from the bodies and the wastes of the crusty lichens, and the dust has been wetted with a rare rain, you may see a whole row of these foliose lichens. The leafy shapes are big and soft enough to catch wind-blown dust. They also contribute bigger bodies when they die and more secreted wastes to enrich the soil in that crack. The spores of more complicated lichens come along with the dust on the wind and find they can live and grow here. Where the crack runs around to the shaded north side of the rock, the adventurous spores of a moss may take hold. Behold, by now there's enough stuff for a grass seed to catch hold, swell up, send down a root; so here comes a blade of grass!

There's a different lichen, common in the desert areas, that isn't on the rocks at all. It lies loosely on the surface of the soil, a gray-green plant smaller than your little finger, looking like a torn bit of thick paper rolled up, black inside the roll, branched a little, no roots. Some damp September days they are so numerous that the whole earth looks gray-green.

In the north desert of my study area there are no trees except a few small willows and a cottonwood bush or two in the channel of Mason Draw. In the south desert there are three limber pines within my walking circle. They are the remotest outposts of the forest that covers the shoulders of the mountains all around us, but separated from us on all sides by miles of high desert. The three young trees may be signs of change, signs that perhaps the forest is creeping lower, signs of possibly more moisture and more coolness. On the other hand, they may be random adventurers, due to come to a bad end because the greater moisture and coolness they need simply do not come along.

By late May you can smell wild onions. They are in bud, scattered here and there between the sage bushes, the grass not yet tall enough to hide them. They will open early in June, flinging up little creamy-white to pinkish bunches of lilies, arranged in a loose flat-topped head, the kind the books call an umbel.

Much later another umbel flower-head makes a yellow glow on the top of the alluvial fan. The sulfur buckwheat comes to bloom in mid-July and is still presenting fresh flowers in September. Where it thrives there's not much of either grass or sage. There are lots of small tumbled rocks, a lot of sand, and the wild short dry channels of the last activities of the last big storm. It's a close relative of domestic buckwheat, and chipmunks love it. I see them eating the flowers and in late fall the fruit. The round yellow flower-heads look at first glance like chrysanthemum pompoms raised on eight-inch stems above a silvery-green hooked rug. On a closer look, the hooked rug becomes a mat of thick oval leaves, each white-wooly underneath and smooth green on top, and each pom-pom becomes a bouquet of yellow flowers on slender stems, all tied together at the top of the main flower-stem with a knot of small green-silver leaves.

Just before the sulphur buckwheat blooms on the rocky top of the alluvial fan, the same area is brightened by masses of a smaller yellow flower, stonecrop, starting to bloom the last week in June. Its fat little leaves aren't really green—say greenish or reddish or purplish. Packed into cracks in big rocks or among the stony gravels, they look like mossy tufts. But the flowers tell you differently. Masses of bright yellow five-pointed stars hide the fleshy leaves, make a ground-hugging golden carpet. The function of the succulent leaves is to treasure the scanty moisture for the sake of the flowers; after a couple of weeks of bloom, when the flowers fade, the whole plant dries up and almost disappears (if you know what you are looking for and go down on your knees, you can find the dry stiff red stems). This is a perennial, keeping life in its roots year after year. It may not bloom at all this year if the weather is dry. It will survive several years without blooming. But the next rainy spring—here comes that golden carpet in the unlikeliest spot.

The pea family is well represented in my desert space, mostly by members of two genera called *Astragalus* and *Aragallus* (or maybe *Oxytropis*, depending on which book you are reading). In a way these pretty little plants are even more frustrating than the willows, grasses, or mustards. They are so showy in bloomtime, these little mat-forming bouquets. The flowers look familiar, like miniature sweet peas, and come in colors as varied as sweet pea's: blue-violet, pink, magenta, purple, red, red-violet, violet, white, yellow. There is great variety in leaf style. They are so conspicuously different that it should be the easiest thing in the world to run them down in a flower book (I thought). How wrong can one be? *The Plants of Yellowstone Park* says: "The following 18 species of *Astragalus* have been identified in the Park"—and then lists eighteen Latin names. It lists eight of the *Aragallus* (or maybe *Oxytropis*?) genus and adds the information that some of these are poisonous to cattle and sheep and produce a disease sometimes called "staggers." The Craigheads' *Field Guide to Rocky Mountain Wildflowers* says that "there are about 1500 species of *Astragalus*; perhaps 100 occur in the Rocky Mt. area. Some species are poisonous or seleniferous; others are among the best domestic sheep feeds. Although the roots, pods, and peas of quite a few species of *Astragalus* were eaten by Indians of various tribes it is not recommended that they be tried by the novice."

With all this vagueness of description, it seems the best I can do to make known my particular friends among the vetches is to include a page of drawings.

It's a pleasure to turn from these confusing plants to some that there's no doubt about. The scarlet globe mallow is one. Hollyhock is a mallow, and all mallows look something like hollyhocks. I hear this one called desert hollyhock sometimes, and that's a better name than the books give it. The mallow part I will take, for it is a hollyhock, but where is the globe? And the color is orange not scarlet. It blooms from May into July in very dry places, particularly along the highway but sometimes on the alluvial fan. You just can't ignore those little orange hollyhocks, sending up masses of shin-high spikes of blooms.

Mentzelia or blazing star is an astonishing plant. Early on a July morning I came upon it in Leseberg Ravine on the west edge of the study area. The slope was so steep and rocky that you'd not think such big plant could grow there. It stood almost hip-high, stocky, sure-footed, and out of its top grew the fanciest big pale

yellow flowers. I'd want to call it blazing star even before reading the book name. There is a great mass of long yellow stamens shooting out above the widespread pointed petals—about 200 stamens per flower. I was lucky to have found it so early in the day, because its bloomtime is at night, and it closes up tight once the day gets warm; I'd not have noticed it then.

The prickly pear cactus is scattered all over the desert, most of it on the alluvial fan and on the foothill slopes of the Badlands. It grows only a few inches tall but spreads its palm-sized paddles over a space of several square feet. These green-gray waxy paddles, they tell me, aren't leaves, but flattened fleshy stems. A single one poking out of the ground by itself is a plant, but most often there are quite a few linked together; or you may see a single normal-sized one with several little ones, like thumbs, growing out along its edge. Long barbed spines (sometimes) and lots of fine bristles (always) grow out of little pockets scattered over the surface. These are loosely attached to the plant and may easily become quite firmly attached to you. They present a particular hazard to fence crawlers. You quickly learn to avoid them. There are 250 species of prickly pear, but I've found only one in my area, the plains cactus. After a winter the pads are flabby, wrinkled, and gray as rhinoceros hide. Soaked with the rains and snow of May and warmed by the high sun, they swell, smooth out, and turn green. In June fat buds as big as your thumb appear in the joints and along the edges of the pads. They grow slowly, and slowly the delicate peach-colored petals unwrap. Then one bright morning in late June, one incredible morning, all over the desert the cactus is in bloom. The flowers are exquisite, the size and shape of teacups, the glowing waxy petals colored light yellow, greenish yellow, lemon, cream, and yellow-orange. For all their delicacy, they are long-lasting; it's a couple of weeks before they fade. They must be very sweet, for the honeybees and bumblebees and ants and beetles love them.

I wonder if the plant has a scent. The flowers have a light fragrance for a short while; but my blind dog Buttons threads her way among cactus plants all year long with never a misstep. Barbed wire scares her; she is very careful and tentative when she is near a fence; sometimes she stops halfway through and asks me to help her; but cactus never makes her hesitate. I've tried myself to smell it, but it's too elusive for me.

The rock rose belongs to the evening primrose family. Long shining leaves spread on the ground in a disk, and on the cluster as on a plate are some great big flowers, big as the palm of your hand and curved like a shallow cup, white or pink or red, with a sweet fragrance. It takes several days' watching before you realize that all the colors belong to the same blossom. It opens up in the night as white. It stays open through the daylight hours, gradually turning pink; just before it wrinkles up it turns deep pink or even red. There are four very wide petals, each with a double scalloped edge. A strange thing about rock rose is that it doesn't have a stem. The thing that looks like a stem between the flower and the ground isn't one at all. It's the calyx-tube of the flower, between petals and ovary, not much different from the throat of an Easter lily, and the seeds develop right at ground level at the bottom of this tube. When the plant quits blooming, withers, and dies down to the ground about mid-July, a close relative takes its place. The white-to-pink flowers look much the same but not quite as big, and they grow sideways out of a white stem a foot tall. The ovaries are in the shelter of the leaves growing up the stem, and they are separated from the petals by a calyx-tube like rock rose's. This is the pallid primrose.

The big showpiece of the spring carpet of bloom is the carpet phlox. Everyone who knows the phlox of old-fashioned flower gardens will immediately recognize the blooms and the fresh sweet phlox scent but will be amazed to see the white and pale-blue masses right on the ground. If you sit down in the basket of bloom and slide your hand under a plant to lift it up, you can see that the flowers and a lot of narrow green leaves grow from a crooked woody stem a foot long that crawls on its belly among the rocks.

Miner's candle makes a scattered white accent over the carpet. It's a white forget-me-not growing in the shape of a candle. Each little flower on the spike has five white petals and a yellow center. In the carpet, too, are more typical blue forget-me-nots, but the flowers of this species are too small to show up much. This one is called stickseed, because the little hooks on the seeds hang

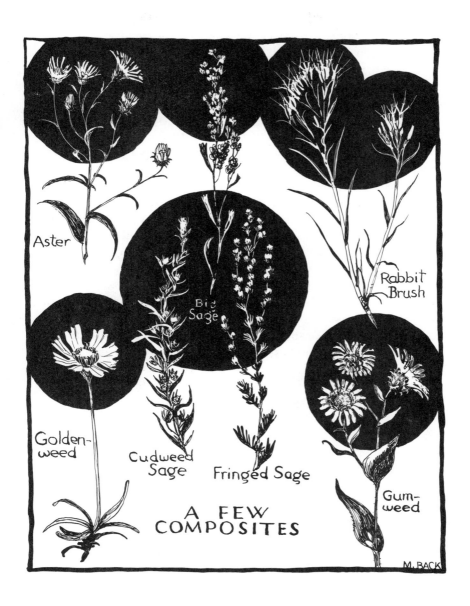

onto your clothes, so people are aware of the plant though they never see the flowers.

The rest of the common desert flowers belong to the composite family; one more of the toughies like vetches and grasses and willows. Each flower of any composite is a whole bouquet made up of a large number of tiny blooms. Typical is the sunflower. Another is sage. You may live in Paris or India or Japan and have never been near the American West; even so, you will know that it's cattle country and a land of grass and sagebrush. This is true.

Sage's fragrance, like a mixture of turpentine, ephedrine, menthol, and camphor, evokes the spirit of the West. It's so powerful that once you've known it you can never forget it, but it's so pervasive that when you've lived here awhile you can't smell it any more unless you crush the narrow gray leaves in your fingers. Almost everywhere it grows in company but alone. You don't get tangled in a sagebrush forest; you can walk in any direction through it and around each bush. A mature bush will be from one to twelve feet tall, generally about three feet, and tough and woody. The yellow to dark gray bark sheds off in long strings, which hang about the branches and litter the ground. Parts of almost every bush are dead wood, tough enough to stay in place for years. Each leaf is about an inch long and a third as wide, the outer end cut by three teeth. Leaves are gray and furry, only faintly green. But they are evergreen; the bushes are never bare; the faint green color is a hopeful sign in the depth of winter; in a wet June the brush is almost real green, graying again as summer wears on.

The wood is full of fat and burns well. Even green leaves are fat and will burn. In much of the West, sagebrush is the only wood for fuel, but here we are so close to larger timber that sagebrush is seldom used; pine, spruce, fir, and cottonwood make much longer lasting, much more easily cut and stacked fuel. Sage is beautiful wood; very big pieces may be six inches across and a crooked six feet long. Woodworkers like it for fancy inlay pieces. It has a tight grain, often with contrasting colors ranging from lemon yellow to purplish brown, and it takes a high polish.

This is the big sage, by far the most abundant. But there are some twenty varieties of sage in the Rockies, and two others are common in my study area. Mat sage, cushion sage, fringed sage is a silver-white plant with straight and narrow leaves close to the ground. Each plant is a flat pad a few inches across, with its little leaves reaching straight up. At bloom time gray slender shoots grow up nine to twelve inches and bear many little yellow flower heads.

Another sage is the cudweed. This is a round little plant with soft stems spraying out from a common center and growing half a foot high. It has yellowish flower heads hanging from each stem and small linear gray leaves an inch or two long. I am told that domestic sheep browse it with enthusiasm, hence its name. In our part of the valley there are no domestic sheep, and I've never seen the bighorns eating cudweed. Deer are common and browse the sages.

There are strong and violent arguments about the place of sage in our economy and environment. There are those who would get rid of it—burn it off and poison it off—in the interest of more grass for cattle. Sometimes it's cattle ranchers, sometimes the U.S. Forest Service, sometimes the Bureau of Land Management. Their line of argument I've suggested already: before the arrival of the white man sagebrush was a negligible factor in the environment; it has increased as grass was killed out by overgrazing; and to bring grass back sage must now be killed.

But such a procedure does violence to the whole present plant and animal community. Sage is big enough and branchy enough to trap the scanty snowfall and hold it until it can melt and make usable moisture, as grass cannot do. It is big enough to make shelter in severe weather for deer, rabbits, grouse, and other wildlife. Wherever it is not overgrazed, grass grows abundantly between the sage bushes, and a great variety of other plants, many of them good forage, feels welcome here, too. If the sage is killed and the ground bared off, then seeded just to grass, the snow blows off leaving brown and blowing earth without enough moisture to keep the grass crop good, and all the shelter for wildlife is gone. Besides, sage itself is food. Deer grow fat on it; some local hunters prefer to hunt deer higher, in the timber, because "if you kill deer at our elevation you can smell and taste the sage." Others like the flavor (though it is bitter, not like garden sage, a quite different plant). Rabbits eat quite a lot of it. It is almost the sole food of the sage grouse. The evergreen leaves are available when snow covers the grass. The pocket gopher my neighbor brought me had his furry cheek pockets stuffed with the three-toothed leaves of big sage.

Rabbitbrush is common in parts of my area, more on the north side than the south. It grows as separate bushes, scattered like big sage and looking something like it. Its stiff woody branches grow to a dome shape, symmetrical and not crooked like the sage. From late July through most of September the dome is plated with gold, bright golden composite flowers massed much like goldenrod. When many bushes are near together the effect is spectacular. Land managers tell me that big sage grows where the soil is pretty good, but that rabbitbrush is a sign of poor or misused soil. It would seem that most of the land in the desert part of my study area is pretty good, because it grows sagebrush, but that our acreage and that of our neighbors' just north is pretty poor, because that's where most of the rabbitbrush grows. Well, it's pretty anyway, and the rabbits like it too.

Here and there in the sagebrush country is a big hole. It's wider than your thigh, and the bottom is out of sight in blackness. It's a badger hole, most likely where a badger has dug for a picketpin, chipmunk, or mouse, rather than for a home. These holes are dangerous, especially where cattle are being handled with horses, for a horse, with his eye on a particular cow, may not watch his footing closely enough and may stick his foot down a badger hole. Often in such a case the horse breaks his leg and must be destroyed. Or the rider, thrown as the horse cartwheels, may break his back. So badgers for this reason alone are persecuted animals. I've not often seen them. There are always a lot more holes than badgers. One time Buttons was barking with such excitement that I couldn't get her attention at all and followed the sound to see what happened. Up on the bench above the highway she had surprised a half-grown badger, who backed up against a sagebrush and stood his ground. A broad, flat, heavy-bodied, yellowish gray creature, he was much smaller than Buttons, maybe twenty inches or so long, counting his shortish tail; but you and she could see he'd be more than a match for her, and Button's frantic barks were coming from a respectable distance away. True to his reputation as a great digger, badger was digging himself in all the time he was challenging Buttons. A flick of his eye gathered me into his consciousness, but he missed not a beat of digging nor a breath of threatening sound. What awful sounds! Combining growl, snarl, whine, and yell, they were scary when he exhaled, but terrifying when he inhaled. He was digging with all four feet in an explosion of sandy dust, and the earth was halfway up his sides when I came on the scene. Buttons couldn't hear me even at close range, so I made my bandanna into an impromptu leash tied to the ring in her collar. Before I could get her dragged away, badger was underground except for his face, mainly baleful eyes, open mouth, and shining sharp teeth; and the noise was more awful than ever.

Of all the weasel tribe, badger is outstanding for poise and courage. He is intelligent and adaptable and can accommodate himself to all sorts of situations. As pets, I am told they are as clever, domestic, and affectionate as dogs. As wild creatures, they make their own way with little contact with people. The holes they dig are the only things man has against them.

Coyotes are diggers, too. Like badgers, they prey on such hole-digging rodents as prairie dogs, picketpins, pocket gophers, and mice, but they have neither the persistence nor the digging equipment of badgers. Digging is seemingly a fun operation with them and if it turns into drudgery, it's time to stop.

Coyote compares favorably with man as a versatile creature. He lives and lives well in the desert, in cultivated farmland, in the suburbs of cities, in mid-western woodlots, and in high western forests. He ranges from the Arctic coast of Alaska into Central America, and from Vermont west to the Pacific coast. I may see him or find his sign anywhere in my study area, but mostly in the desert. I think only one or two families range into my area. I have never found a den, but dens there surely must be since coyotes continue to appear. I have no information on such birth-control measures as have been found to be practiced by wolves to keep the population at the level of the food supply, but I suspect they exist.

Buttons has grown sophisticated with her years and can no longer be lured into games with coyotes, but their near presence still excites her. Once in a while young ones try to play with her, but most of our coyote-watching is from a distance, and most of our encounters are with their songs. I read that much of their yapping is designed to keep members of a family in touch with each other when hunting on dark nights, but surely part of it is pure singing for the joy of producing and listening to the sounds. Sometimes you can identify several members of a family—father, mother, several pups, maybe an aunt or uncle or even grandad—sitting on separated lookout points, enjoying the blended harmonies pouring from their throats.

The ranks of the coyotes have been somewhat reduced and their family structure disoriented by the dreadful and long continued poisoning campaigns initiated by stockmen's associations and fostered for years by the Federal Government—mostly that branch called curiously the "Fish and Wildlife Service." Very many other creatures were killed by the coyote poison, including large numbers of dogs and eagles. It looks as if the tremendous killing has set up some compensatory breeding mechanism among coyotes, because the total population around here hasn't seemed to change much. In my little study area there hasn't been any great amount of poisoning, but I still rejoiced when the executive order stopped the practice on all federal lands. I am thankful every time I hear the choruses in the night.

What mainly keeps down the population of coyotes and badgers is the comparative lack of their chief food, rabbits and small rodents. This should be great prairie dog country and used to be. A poison campaign to eradicate them was carried on for years by the Biological Survey and its successor the Fish and Wildlife Service with only too much success. I thought that in my study area they were all gone, then six years ago on a low sandy ridge between our house and the Badlands, I watched a solemn conference of half a dozen little brown fatties that turned out to be white-tailed prairie dogs. Maybe they were an advance committee of real estate developers—it did work out that way. Holes and mounds appeared, with fat golden householders sitting on the mounds, and ducking into holes when a shrill alarm whistle sounded. I watched them with great interest, which turned into grief when I found they had been selected as targets by local shooters. Now for three years

THE DESERT

I have seen no prairie dogs. But a recent walk over that sandy ridge showed me some fresh-looking holes by some fresh-looking mounds.

The picketpins or Uinta ground squirrels are the staple food of the coyotes; their neat small holes are all over the desert, and their squashed bodies on the highway attract clusters of magpies and ravens. Right now we seem to be at the bottom of their population curve. They are pretty scarce.

Chipmunks were scarce last year but abundant this year. They love the sage as well as the thickets, nesting and growing up in its shelter, eating its leaves, climbing it for lookout posts or to snooze in the sun on its thickest foliage. When the whole desert community is destroyed by campaigns to wipe out sagebrush and replace it with stands of grass, the chipmunk population is wiped out with it. They are decorative, picturesque, friendly little fellows, and fill an important economic spot as well, being food for the coyotes and the big hawks.

Jackrabbits are very scarce. In the years of my study they have never been abundant. The ones that live here are the white-tailed jackrabbits, who change color and get white or pale gray in winter. That's dangerous business in our part of the valley, where snow is generally scanty or absent. That fact alone might account for their scarcity.

The little cottontails, the bunnies, here as almost everywhere are the staple food of the predators. They keep up their ranks by their sheer power to multiply. They live in the woods, the thickets, the edges of the hayfields, the ditchrows, the fencerows, and the desert. I see most of my cottontails in the desert.

I look for four species of ruminants, the large game animals that chew their cud: mule deer, moose, pronghorns, and Rocky Mountain bighorn sheep. Of these, mule deer are common residents, moose occasional winter residents, and pronghorns rare. Bighorn sheep are uncommon but increasing, especially in the Badlands.

Though you see mule deer in the thickets, the sloughs, and the woods, their home is the desert. If you are hunting for deer, you go to dry hills with scattered cedars or limber pines or just out on the sage benches. I expect to see them any time of the year.

They are usually in groups of three or four, most likely a couple of does with their fawns. Often they gather into small herds of a dozen or so. They are gentle animals with large dark eyes, much larger than the eastern white-tail; a mature doe, dressed out, cut up, and wrapped for the freezer, will weigh about a hundred pounds of meat, a big buck up to 200 pounds. In spring and summer they are brown, with white faces, throats, inside ears, and rear ends, accented with black in eyes, ear edges, muzzle pattern, neck and breast, hoofs, and end of tail. In fall and winter the white and black patterns are the same, but the brown part changes to blue-gray. Their ears are very large, whence the name mule deer. When they look at you with interest the ears flare out widely, and they move them back and forth as you watch. When they run, they fold them and hold them high the way jackrabbits do. I don't often see bucks, they are more timid than does. I look for the bucks, for a big one is a magnificent sight. He carries himself proudly, and the ivory rack of horns is worth the pride. The antlers sweep backward; each divides into two Ys that reach upward, then divide again into two Ys.

The fawns are born on the desert hills in the shelter of sagebrush. It seems to be really true that the young have no odor. Mother will apparently direct a fawn to stay curled up and be still whatever happens, then she will go off, maybe a mile or so, and feed, far enough away so a marauding coyote will have no hint of the direction of the sleeping fawn. Once a rather large party of us came upon a very young fawn curled up in a narrow draw draining from the moraine into CM Draw. My friend Fran Primrose, her two small daughters, and my two dogs were with me. If we hadn't almost stepped on the little one we'd not have seen it. It was curled up in a tight circle, legs folded close, chin on thigh, shining eyes wide open. The back was a rust color, dappled with many small white spots. The only movement was a flicker of eyelids. The little girls wanted to pet it, but I warned them not to touch. We looked at it from a circle a few feet away, then went on. It was notable that the dogs, dashing about on doggy business in all directions, never found the baby. We felt sure the mother was a doe we were watching a few minutes earlier and a few hundred feet away.

THE DESERT

Deer are abundant enough for us to feel that they are a crop to be harvested, like wheat or beef. We know that they do not practice birth control, as wolves and perhaps coyotes do, and for that reason are subject to population explosions and starvation die-offs. Deer meat is excellent eating. When we have it, we enjoy it and never waste it; we freeze it and use it thankfully till it is all gone. But deer are so beautiful and gentle that our hearts do fail us, hunting them.

We made a pact not to shoot deer near home—let's go miles away and shoot strangers. We do not justify the killing, beyond citing the facts that every bite of any kind of food we eat was part of some living creature who gave up his life for us and that like all other living creatures we can survive in no other way than by taking life.

One of the core experiences of my life is tied to a deer hunt. I tell myself that it was a hallucination born of fatigue, but I am deeply aware that it was more than that. At the very least it was a powerful insight into a mystery. Joe shot a doe a few miles from home and a mile from our pickup. It was nearly dark and haste was indicated. While Joe dressed her out, I raced down the draw to the truck, got rope and tarp, climbed back with them up the steep rocky draw, and helped Joe drag, boost, and carry her the long mile down. While Joe moved things in the pickup to make room for her, she and I were side by side, waiting in the frosty twilight. She lay quietly on her side, her insides gone and her eyes dusty. Suddenly she spoke! She said, very clearly, "This is my body which is given for you. Take it and eat it. Do this in remembrance of me."

Certainly I have remembered. Every meal since then has been a communion with all of life, the whole body of which I am a part.

Moose are never listed as mammals of the desert. My usual meetings with them are the winter ones I've reported in the chapter on Wind River. But moose are the least predictable of the creatures I know. I'll never forget one incident, when a lady tourist asked me about where to see moose, as we stood near the highway under a warm July sun. I answered her as best I could: "Look for them in the willow thickets thirty miles and more ahead of you.

You never see them this low in the summer." Her comeback was "What's that?" And sure enough it was a moose! We saw her on the skyline just above us, where the Low Bench breaks from a flat to the sharp hill down to the highway. She made her way down the slope through the boulders and sagebrush, jumped the highway fence, made cars stop as she crossed the highway, jumped the other fence, trotted down to and through the river, and with a long swinging trot went toward the Badlands. We saw it, though we know it can't be so.

The desert is real pronghorn antelope country, but in my study area I must call them very rare. When I came to the Wind River Valley almost fifty years ago they had been exterminated here. Strict protection on both state and federal levels had allowed them to make such a great comeback that they are now among Wyoming's most valued game animals. In our valley they have returned from complete extinction. There are no fences at all in the Diamond A pasture, a small part of which is my north desert, and antelope range freely in it a few miles northwest. Sometimes I do see them.

The pronghorns are strange animals, so unlike any others that they are given not only a unique species, but a unique genus and even a unique family. They are said to be closer to goats than to any other family, and "goats" they are often called; but they're really not much like goats. An antelope is small, hardly forty pounds of usable meat, and built for speed. They've proven themselves able to run long distances at forty miles an hour. They're long-legged and short-backed and remarkably clean and slender of limb. All the other ruminants except giraffes have dew claws, a pair of small hoofs that don't usually touch the ground, above and behind the functioning hoofs, but the antelope have no trace of them. Their color is light tan above and white below, with a sharp line of demarcation and no blending. There are two tan rings around the neck, one or both of them broken by white. There's a very large white target around the short dark tail. They can at will stand the white hairs here on end, flashing out a signal like turning on a light. They have extremely large dark eyes set well out from the sides of their heads, and the black prong horns, hooked like shepherd's crooks, grow right up above the eye sockets. These are the most curious horns. Once I stewed a horn for twenty-four hours, having been told it was an interesting stunt. It completely disintegrated into a mass of black hairs and a pot of glue, leaving me with the mystery of how it all grew together in the first place.

Another mystery is why I have seldom seen bighorn sheep in my study area. These are the rarest of North American game animals, so not many students would find it strange to be denied them. But only a few miles away and a few hundred feet higher up, the Whiskey Mountain herd ranges free and unafraid; this is the largest herd of Rocky Mountain bighorns there is. On Sheep Ridge they winter by the hundred, and from Ramshorn Street in Dubois people watch them through binoculars; the hill above the pole gate at the top of my alluvial fan is the lower end of Sheep Ridge.

Five species of hawks are sage country predators. Red tails give it some attention in spring and fall, rough-legged hawks in winter. Bald and golden eagles happen over it once in a while, and I've seen prairie falcons all year except deep winter. None of them, I think, have nested in my study area during the thirty years I've been watching it.

A pair of red tails shows up every year in late March or early April, stays for occasional sightings through May, then disappears to show up again in late August or September. I've two records in June and one in July. I surmise they nest not very far away, but not here. Typically I see them soaring, showing the red tail as they bank. A few times they have treated me to a good close look as they sat on power-line poles and preened. They are magnificent birds, powerful and fierce, and with rich, soft, lovely plumage. The sleek feathers of the back and outside of the wings are dark, the tail bright rust edged with black and gold; the thick satiny masses of the breast and thigh feathers and the wing linings are light cinnamon flecked with dark. There are white accents around the face and the inside of the flight feathers. I wish they would stay with us longer and let us see more of them.

I'll not forget a courting flight soon. The two birds staged the display just south of the Badlands cliffs. Both of them screamed again and again as they flew around each other in such small circles that I couldn't believe it. They kept it up for all of fifteen minutes. Then the circles increased in size, till the two birds, still soaring in interlocking circles, disappeared beyond the high red ridge.

One of the few chances I've had to witness a hawk's success was given to me by a red tail, the first sighting of the year, on March 25. When I saw him, he was dropping out of the sky where the moraine is near the river. Near the ground he flattened out and flew very fast across the highway, no higher above it than the top of a car, made a spectacular high speed side-slip into sagebrush on a steep rocky hill, and neatly caught a rabbit.

Rough-legged hawks are winter birds. I have seen them in every month but June and July and quite regularly from November to March. They are about the size of redtails (wingspread about four and a half feet), but seem a little slower and heavier. They often fly very low, and one may hover with beating wings over the same spot for several seconds at a time. (I suppose he saw a mouse but wasn't sure of his aim.) Often a roughleg is misidentified as a bald eagle because his head may be mostly white, and his tail has a lot of white on it. But when the head is white most of the breast is too, which is quite unlike the dark breast of the eagle. Several times I've thought "bald eagle" when it was a white-headed roughleg sitting on top of a power-line pole with his back to me. But the head is not totally white; it is flecked with dark, and there is usually a dark area from the eye back toward the dark neck feathers; and the tail is not all white, just the half nearest the body. Roughlegs come in a confusing variety of plumage pattern, even to one that is almost black.

Bald and golden eagles are sometimes seen over the two desert areas. They are accidents, I guess, for the bald eagle haunts the river, and the golden likes cliffs. Just once I saw eagles on the ground in CM Draw. These were two goldens sitting side by side, big and black, on a granite boulder on the edge of the moraine. I gathered this was a courting sequence. One partly raised a wing and nudged the other with his elbow. She gave a shudder, raised her shoulders, sank her head between them, and gazed impassively across the draw. Across CM Draw. And saw me! She gave a bigger

shudder, swung her head around to her pal, and gestured with it toward me. The first eagle came out of his bemused trance and saw me too. Both eagles tried to take off from that boulder at once, but there wasn't room. They hit each other with their wings, and one fell off the rock, spreading wings and landing in a heap. The other shook out her feathers, then deliberately took off into the west wind, across the draw, past me and gaining height, over the Low Bench. When I looked back, the one left had regained the boulder and was in the act of taking off. I was sorry I'd interrupted what might have been a touching love scene. Ah, well!

The prairie falcon is a sometime visitor, too. No such clumsiness with the falcon as with the eagle. Prairie is a trim, powerful, predatory machine. I have sixteen sightings listed, spread over eight years and in the months from April through December. This implies that they nest somewhere near, though the small number of sightings indicates their visits to my area are accidental. The prairie falcon is the size of a raven. You recognise it as a falcon the moment you see it. Its long pointed wings and closed tail are signs; even more is its resemblance in flight to the much smaller kestrel. It flies very fast, makes you think of a bullet, but can stop and hover easily over one spot. Also it can swoop from high in the sky almost to the ground to catch its prey. The most important field mark is black armpits, very conspicuous in the light underside of the bird as it flies overhead.

Colored like an owl, about the size of a robin, and like nothing but himself, is the nighthawk, who needs the woods for resting. Yes, I said resting, not nesting. He nests in the desert. A nighthawk's body is about the size of a lemon, and most of the lemon is breast-muscle, the engine of flying. It has a medium-sized tail, tiny and almost useless feet, and long narrow wings: a bird eight and a half inches long may have the incredible wingspread of twenty-one inches. Its head is almost as large as its body; seen at close range it looks all eyes and mouth. The mouth splits literally from ear to ear, a gape like a frog's, and almost as soft as human lips. There is a tiny hardening in the middle of each lip, so when the mouth is closed you can see a little dark beak. This anatomy is covered with a coat of owl-soft feathers in many tones of gray-brown. There's a pure white bar across the middle of each wing and a triangular white throat under the wide mouth.

Long before I knew about nighthawks' need for the woods, I'd admired the birds as flyers. From our picnic spot by the river we've watched them fly great arcs across the sky. They are somewhat social, half a dozen or so at a time making an aerobatic display. One may fly with long easy wingbeats for a while, shift into a phase of quick flutters, coast some, then back to a smooth long rhythm. They are experts at instant and extreme changes of direction, at near misses, at weaving dissolving patterns in the sky. The climax of the act is the BOOM. In turn, one after another will fold wings and drop with heart-stopping speed straight for the ground. Almost by your ear he will level off, every wing-feather thrumming into a total of BOOM! Then he coasts back up into the sky, using left-over momentum and no wing beats, on a long slant. His other sound is a plaintive single note, "Peent!" He may cry it out into the sky as an accent to the visual pattern of flight, or he may utter it at intervals from his resting place in the woods.

This aerial patterning is no circus act to thrill human spectators but is part of the serious business of life—eating. Nighthawks live on flying insects taken from the air. I have seen nighthawks take mosquitos, butterflies, and moths at my level, but I have no idea what it is they look for so high in the sky. It looks as if there must be an over-heavy outpouring of energy compared to the amount of food taken. But the continued existence proves that the system works. However rich the food may be, the chase is certainly exhausting. Nighthawks rest most of the time—almost always all night and most of the day. They usually fly in the twilight of early morning and late evening, occasionally in the full light of day, and sometimes by the light of a full moon. Makes you think that mosquitos and other insects represent a source of high-energy food that our human diets should know more about. Piece of mosquito-pie, anyone?

The long rests are taken in the woods. I found my first resting nighthawk when I heard a loud "Peent!" just over my head. He was playing games. I looked long and carefully in vain. I gave up and started on when "Peent!" came again, loud and close. This time I found him, lying flat and lengthwise on a branch ten feet above me, looking like an extra heavy piece of bark. A big dark eye was wide open and round, looking at me. The game was over by then from his point of view. The round eye narrowed to a slit, eyelids like shreds of gray bark came together, and I guess he went to sleep. The next day he was in the same place, and the next, all summer long. Once he had taught me the secret, I practiced finding resting nighthawks.

The nighthawks that rest in the woods nest in the desert. Because I live in the woods I see them more often there. I've learned to know the branches where they roost and look for them every summer day. I know how they fly rapidly in through the curtain of leaves, make a perfect landing, fold wings, and are at rest, all in one motion. In contrast, I've only twelve records of scaring them up off the ground in the desert, and I've never found a nest. Nest? Peterson says no nest at all, just bare sand or gravel. Each time I've seen one in the desert it's been just after I almost set my foot on it, so each time I've looked for eggs. If they sit that close, I could easily have missed a lot of nests just a foot or two to the side of my path. Several of the ones I've flushed in July and August looked to be juveniles not sure of their flying. One seemed quite inept; seemed he could barely clear the sagebrush, but he did keep himself somehow in the air for several hundred feet before he pitched into the sage again.

Horned larks are built for the open country. I often see small groups of a dozen or so in the north or south desert. I'll hear them first, a high sweet twittering. Following the sound, I'll finally see them, maybe high in the air and pitching down, maybe low over the sagebrush. They don't seem to be tied to territories, they wander freely and erratically over any open country. They tend to bunch up in stormy weather. One dark April morning I found myself in the midst of about 500 in the gravel pit on the Low Bench. The air was full of movement and so was the ground—little brown-and-yellow mouselike birds scooting about, walking, running, and flying. They don't hop. They were eating seeds of ragweed and sweet clover. A dozen or a hundred would rise at once, fly twenty feet or so and settle down, displacing another group who would do the same. It was a lively and contagious action—I felt like flying up, twittering, and settling down with them and so did the dogs!

In the red cliffs of the Badlands, cliff swallows and violet green swallows nest. The cliff swallows build sturdy nests out of mud on ledges and vertical walls of the cliffs. On May 30 I watched them scooping up in their mouths little pellets of mud which they stuck with tapping pressures of head and neck against the wall. It takes several days to build the adobe apartment. When it's done it's roughly ball- or jug-shaped, with a swallow-sized hole in the side. Sometimes dozens of these will be side by side and above each other for several ranks: genuine apartment houses. Cliff swallows are bright-colored little birds, light below with dark red throats, shiny dark above with square black tails, and pale orange rumps and foreheads. Peterson says cliff swallows winter in Brazil, Chile, and Argentina; they have a long way to fly.

The violet-green swallows winter in Mexico, not nearly so far away, yet their dates of arrival and departure are about the same as cliff swallows. They nest in holes and crannies, in trees, and around houses as well as in the cliffs. I've never found a Badlands nest, but the sight of adults flying round the pinnacles in nesting seasons seems evidence the nests are there.

One pair of ravens nests on the nearest Badlands cliff. I see them nearly every time I go in that direction in summer, but the nest itself is so completely out of reach that I've never seen it; I just hear the squawks of the young ravens.

The chance of seeing one of those busy mountain gnomes, the rock wrens, brightens my summer days. They are much bigger than house wrens, almost as big as house sparrows, and are the grayest of all the wrens—darkish above, lighter below, with light eye stripes and light tail corners. Rock wrens and Bewick's wrens are close to the same size and patterned alike (Bewick's are browner above and whiter below), but I never see them anywhere near the same places. Bewick's have been in thickets by the river (only in migration), rock wrens farthest from water in the driest possible locations (and all summer long). They never try to conceal themselves and usually make a point of announcing their presence. I expect to find them sky-lined, on highest pinnacles or nearest fence posts. They are curious and do their share of people-watching when they get a chance.

The main bird of the sage lands is the sage thrasher. He is bold, impertinent, and everywhere. He has a loud, clear, musical voice and a long varied song, and his territory is as much sage land as is in reach of his voice from his nest. So his territory is more than a quarter of a mile across, less upwind (west) and more downwind (east). Other sage thrashers are likely to have territories touching this one in each of several directions. You can't sneak up on a sage thrasher—he can see you coming a mile away. If he can't keep trespassers off with his voice, he'll fly at them at close range.

Though closely related, he's a different character from the lurking catbird, the calm brown thrasher, or the aristocratic mockingbird. He is inconspicuous in appearance but not in action.

It was fun to watch a sage thrasher doing a great job of grooming. He concentrated so on making himself beautiful that he failed to see me approaching. He was turning a wing inside out when the flash of light color caught my eye so I turned the field glasses on him. He sat on a fence post in CM Draw, polishing each inside wing-feather and pulling it into position with a little jerk. Satisfied at last, he refolded the wing properly, shook the whole thing into place, and turned the other inside out. That finished, he stuck his left leg out in front, balancing on the other while he straightened each feather; he transferred his balance and did the other leg the same way. He pooched out his chest, doubled his neck up, and ran his bill several times down the fluffed-out breast feathers. He shook himself all over, and looked beautiful. From the date, June 30, I had to assume his youngsters were a-wing and finding their own grasshoppers; maybe it had been a month since the last good grooming. Even more likely, this was Mother instead of Father! Perhaps Father was leading the brood on their first grasshopper hunt, giving Mother her first chance for the toilette I spied on.

Now the first stage of my study of the acres around me has come all the way from the bed of the river, up the banks, through the swamps, underbrush, woods, and houses, to the all-surrounding desert, and the edges of my arbitrary study area. This has been a superficial look. I have uncovered many more mysteries than have been explained. Twenty years of walking those acres has sharpened my need to know and my conviction that we are all woven together into one body, the body of Life.

Somehow, the song of the coyotes, more than bird-song, seems to pull everything together—coyotes, wild and free and so much like us, wailing long and low from the CM pasture in the January dawn, answered by excited yelps from the Jakey's Fork Hills . . . ripping out a jolly chorus in November moonlight . . . pouring forth rich harmony from Mason Draw under a September harvest moon.

In the early moonlit dawn of a February day, Joe and I watched a big dog coyote trot down the hill from the Low Bench into CM Draw and east across it. Shortly afterward Buttons and I walked that way as daylight brightened. I took a good look, and Buttons a good smell, at the tracks in the snow when we crossed them, neat doggy tracks a little bigger than Buttons was making. We'd gone on up the draw when we heard behind us a bark, shifting into a high tremulous wail. We both snapped around and saw the big gray coyote just about where we had crossed his tracks. He was going the other way, back up the hill to the Low Bench. We saw him silhouetted against the Badlands. He stopped and looked at us, one front foot raised. Buttons gave a wild yipe and took out for him, and he floated easily up the skyline. I ran straight up the slope in the slippery light snow, reaching the top edge breathless. There was a picture to see.

That dog coyote had a date on the Low Bench. A slender young female, prettily marked, was waiting for him a hundred yards or so beyond the edge. It was full day by now, the contrast between the two showed up. He was hulking big, broad shouldered, gray. She was grayish tan with delicate yellow feet and yellow eyes, white throat and breast, white trimmings round her eyes and inside her ears, black accents on back and shoulders and the end of her nose. Buttons stood still watching, halfway between me and the coyotes. The big one moved to conceal us from his lady, who had already had her good look. She half jumped up with her front feet, turned around, and trotted off beside her mate across the Low Bench and up the nearest slopes of Sheep Ridge.

Bibliography

GEOLOGY

Keefer, William R. *Geology of the Dunoir Area, Fremont County, Wyoming.* Geological Survey Professional Paper 294-E. Washington, D.C.; U.S. Government Printing Office, 1957.

Love, John David. *Geology Across the Southern Margin of the Absaroka Range, Wyoming.* Geological Society of America Special Papers, 1939.

PLANTS

Beetle, Alan A., and Morton May. *Grasses of Wyoming.* Research Journal 39, Agricultural Experiment Station, University of Wyoming Laramie. January, 1971.

Bohmont, D.W., and H.P. Alley. *Weeds of Wyoming.* Bulletin 325R, Agricultural Experiment Station, University of Wyoming, November 1961.

Craighead, John J., Frank C. Craighead, and Ray J. Davis. *Field Guide to Rocky Mountain Wild Flowers.* Boston: Houghton Mifflin, 1963.

Longyear, Burton O. *Trees and Shrubs of the Rocky Mountain Region.* New York: G.P. Putnam, 1927.

McDougall, W.B., and Herma A. Baggly. *Plants of Yellowstone National Park.* Washington, D.C.: U.S. Government Printing Office, 1936.

Peattie, Donald Culross. *A Natural History of Western Trees.* Boston: Houghton Mifflin, 1950.

Preston, Richard J., Jr. *Rocky Mountain Trees.* 3rd edition, revised. New York: Dover, 1968.

Sudworth, George B. *Forest Trees of the Pacific Slope.* New York: Dover, 1967.

MAMMALS

Burt, William Henry, and Richard Philip Grossenheider. *Field Guide to the Mammals*. Boston: Houghton Mifflin, 1964.

MacClintock, Dorcas. *Squirrels of North America*. New York: Van Nostrand, 1970.

Seton, Ernest Thompson. *Lives of Game Animals*. New York: Doubleday, 1929.

BIRDS

Bailey, Alfred M., and Robert J. Neidrach. *Birds of Colorado*. Denver: Denver Museum of Natural History, 1965.

Forbush, Edward Howe. *Birds of Massachusetts and Other New England States*. Commonwealth of Massachusetts, 1929.

Peterson, Roger Tory. *Field Guide to Western Birds 2nd edition*. Boston: Houghton Mifflin, 1961.

Robbins, Chandler, Bertel Brunn, and Herbert S. Zim. *Birds of North America*. New York: Golden Press, 1966.